Hydrogen, Biomass, and Bioenergy

Integration Pathways for Renewable Energy Applications

Hydrogen and Fuel Cell Primers

Hydrogen, Biomass, and Bioenergy

Integration Pathways for Renewable Energy Applications

Edited by

Jacob J. Lamb

Department of Electronic Systems, Department of Energy and Process Engineering & ENERSENSE, Faculty of Information Technology and Electrical Engineering, Norwegian University of Science and Technology (NTNU), Trondheim, Norway

Bruno G. Pollet

Department of Energy and Process Engineering, ENERSENSE & NTNU Team Hydrogen, Faculty of Engineering, NTNU, Trondheim, Norway

Series editor

Bruno G. Pollet

ELSEVIER

ACADEMIC PRESS

An imprint of Elsevier

Academic Press is an imprint of Elsevier
125 London Wall, London EC2Y 5AS, United Kingdom
525 B Street, Suite 1650, San Diego, CA 92101, United States
50 Hampshire Street, 5th Floor, Cambridge, MA 02139, United States
The Boulevard, Langford Lane, Kidlington, Oxford OX5 1GB, United Kingdom

Notices
Knowledge and best practice in this field are constantly changing. As new research and experience broaden our understanding, changes in research methods, professional practices, or medical treatment may become necessary.

Practitioners and researchers must always rely on their own experience and knowledge in evaluating and using any information, methods, compounds, or experiments described herein. In using such information or methods they should be mindful of their own safety and the safety of others, including parties for whom they have a professional responsibility.

To the fullest extent of the law, neither the Publisher nor the authors, contributors, or editors, assume any liability for any injury and/or damage to persons or property as a matter of products liability, negligence or otherwise, or from any use or operation of any methods, products, instructions, or ideas contained in the material herein.

Library of Congress Cataloging-in-Publication Data
A catalog record for this book is available from the Library of Congress

British Library Cataloguing-in-Publication Data
A catalogue record for this book is available from the British Library

ISBN: 978-0-08-102629-8

For information on all Academic Press publications
visit our website at https://www.elsevier.com/books-and-journals

Publisher: Brian Romer
Acquisitions Editor: Peter Adamson
Editorial Project Manager: Andrae Akeh
Production Project Manager: Sruthi Satheesh
Designer: Victoria Pearson

Typeset by Thomson Digital

Contents

Contributors

Bjørn Austbø
Department of Energy and Process Engineering, ENERSENSE, Faculty of Engineering, NTNU, Trondheim, Norway

Robert Bock
Department of Energy and Process Engineering, ENERSENSE, Faculty of Engineering, NTNU, Trondheim, Norway

Odne S. Burheim
Department of Energy and Process Engineering, ENERSENSE & NTNU Team Hydrogen, Faculty of Engineering, NTNU, Trondheim, Norway

Magne Hillestad
Department of Chemical Engineering, ENERSENSE, Faculty of Natural Sciences, NTNU, Trondheim, Norway

Jacob J. Lamb
Department of Electronic Systems, Department of Energy and Process Engineering, ENERSENSE, Faculty of Information Technology and Electrical Engineering, NTNU, Trondheim, Norway

Kristian M. Lien
Department of Energy and Process Engineering, ENERSENSE, Faculty of Engineering, NTNU, Trondheim, Norway

Anna S.R. Nordgård
Department of Biotechnology and Food Science, Faculty of Natural Sciences, NTNU, Trondheim, Norway

Bruno G. Pollet
Department of Energy and Process Engineering, ENERSENSE & NTNU Team Hydrogen, Faculty of Engineering, NTNU, Trondheim, Norway

Erling Rytter
Department of Chemical Engineering, ENERSENSE, Faculty of Natural Sciences, NTNU, Trondheim, Norway

Shiplu Sarker
Department of Manufacturing and Civil Engineering, ENERSENSE, Faculty of Engineering, NTNU, Trondheim, Norway

Federico Ustolin
Department of Mechanical and Industrial Engineering, Faculty of Engineering, NTNU, Trondheim, Norway

Preface

More than 80% of our present energy consumption is chemical and non-renewable (coal, oil, and gas). The most important renewable energy alternatives, wind and solar energy, produce electric energy that is intermittent, and also inadequate for most transportation options. Bioenergy is currently the renewable energy that is simplest to integrate into the domestic and transport sectors. Moreover, biohydrogen provides a sustainable solution for hydrogen production and is very well-suited to couple with other renewable energy sources.

The intention of this volume is to provide a brief research source for biohydrogen and bioenergy, discussing fundamental aspects as well as cutting-edge trends in (bio)hydrogen and bioenergy production. This volume provides industry professionals, researchers, and students with the most updated review on hydrogen production technology and current trends, thus helping them identify technology gaps, develop new materials, and novel designs that lead to commercially viable hydrogen generation systems.

Dr. Jacob J. Lamb and Prof. Dr. Bruno G. Pollet

Editors of *Hydrogen, Biomass, and Bioenergy—Integration Pathways for Renewable Energy Applications*

[a]*About ENERSENSE*

ENERSENSE

NTNU
Norwegian University of Science and Technology

ENERSENSE is a strategic research area with focus on the nexus of energy efficiency, energy storage, and sensor technologies, including automation.

[b]*About NTNU Team Hydrogen*

NTNU HYDROGEN

NTNU – Trondheim
Norwegian University of
Science and Technology

NTNU
Norwegian University of
Science and Technology

NTNU Team Hydrogen is a team of world experts on Hydrogen Energy. The team consists of researchers from different disciplines, departments, and faculties across NTNU that works within the Hydrogen area. One of the main tasks of the team is to develop new research programs and projects both nationally and internationally with academia, research organizations and industry, and to coordinate Hydrogen activities within NTNU. We are also educating and training research leaders, innovating, providing solutions, and stimulating the industry.

"Conversion into chemical energy (e.g., hydrogen, batteries, and super-capacitors) allows for energy systems that supply energy in the right form, at the right time, and at the right place," Prof. Odne S. Burheim, ENERSENSE leader, NTNU.

Acknowledgments

The editors and authors are grateful to the ENERSENSE program, Team Hydrogen at NTNU, and The Norwegian University of Science and Technology (NTNU) for supporting and facilitating this book project.

Chapter | one

Introduction

Jacob J. Lamb[a], Kristian M. Lien[b], Bruno G. Pollet[c], Odne S. Burheim[c]

[a]*Department of Electronic Systems, Department of Energy and Process Engineering, ENERSENSE, Faculty of Information Technology and Electrical Engineering, NTNU, Trondheim, Norway;* [b]*Department of Energy and Process Engineering, ENERSENSE, Faculty of Engineering, NTNU, Trondheim, Norway;* [c]*Department of Energy and Process Engineering, ENERSENSE & NTNU Team Hydrogen, Faculty of Engineering, NTNU, Trondheim, Norway*

CHAPTER OUTLINE

ABSTRACT

Hydrogen is the most abundant element in the universe. It is the origin of the energy dissipation from stars, by radiation from the nuclear reaction (fusion), where hydrogen is consumed, and helium is produced. On a molar basis, hydrogen is also the most common element on the surface of the Earth. Despite this, molecular hydrogen (H_2) is rarely found on the surface of the Earth. This is due to the abundance of electronegative elements (e.g., oxygen, O_2), which hydrogen has a strong tendency to donate electrons to, forming heteroatomic hydrogen-containing molecules (e.g., water). Water is presently the most abundant hydrogen-containing molecule on Earth, but although water is essential to all life on this planet, energy to sustain life's biological processes cannot be straightforwardly extracted from water. This chapter will introduce the relationship between hydrogen, biomass, and bioenergy.

Hydrogen, Biomass and Bioenergy. http://dx.doi.org/10.1016/B978-0-08-102629-8.00001-3

HYDROGEN, BIOMASS, AND BIOENERGY

Through the radiation from the hydrogen fusion processes occurring in the interior of our Sun, fully oxidized substances such as CO_2 and water can be fixed through biological light absorbing components [1–3], to compounds that can act as chemical energy. The biological mass that embodies this chemical energy is coined as biomass. Biomass is then consumed through the food chain to sustain all biological life on Earth. Additionally, biomass from photosynthetic organisms can become buried through various processes (e.g., sedimentation). Over geological time scales in anaerobic conditions, these biomasses can be chemically transformed into hydrocarbons (e.g., crude oil and natural gas) or coal. These ancient biomass stores hold vast amounts of concentrated energy.

Since the start of the industrial era, buried biomasses have been extensively extracted from under the Earth's surface and combusted with oxygen to release their energy. As a consequence, large quantities of CO_2 have been released into the ecosystem, resulting in excess carbon in the atmosphere and oceans. The accumulation of CO_2 in the atmosphere causes a greenhouse effect, increasing the temperature on the surface of the Earth. As a result, a transition to energy sources that do not increase the carbon content in the ecosystem is urgently required.

Energy storage for the transport sector

Energy sources with low-carbon emissions include solar and wind energy, which are ultimately powered by radiation from the large hydrogen fusion reactor we call The Sun. Moreover, demand in these energy sources has rapidly increased in recent years, yet they are intermittent sources of energy. Since the solar irradiation on the surface of the Earth varies dramatically due to weather, longitude, time of day, and the unpredictability of wind currents at many locations, the reliability of these energy sources is variable. Increasing energy production from these intermittent sources produces a need for an effective method of energy storage.

There are four main types of energy storage that can be readily used for long-term storage. These are electrochemical storage (e.g., batteries), chemical storage (e.g., biomass), synthetically produced substance storage (e.g., alcohols and hydrocarbons), and physical storage (e.g., hydropower dams and compressed air) [4]. A comparison of some of these energy storage systems relative to their weight and volume is shown in Fig. 1.1.

The transport sector requires energy storage systems that have high amounts of energy relative to their volume and mass, and as shown in Fig. 1.1, liquid hydrogen offers the best energy capacity per mass, whereas hydrocarbon energy storage (e.g., diesel, gasoline, and jet fuel) offer the best energy capacity per volume. Liquid biogas and natural gas offer the best compromise between energy capacity per volume and mass, whereas all other energy sources have comparatively low energy capacities per volume and mass. This trend is further exacerbated when looking at the data logarithmically. Moreover, when considering the energy density, relative energy to time release of these storage solutions, both fly wheels and lithium ion batteries (LiB), offer conversion rates that are comparable to that of hydrogen and hydrocarbons, but they still have a significantly lower energy capacity [4].

Molecular hydrogen has long been recognized as a suitable energy storage alternative, but since it does not naturally exist in large quantities in this form on the surface of the Earth, it needs to be produced from hydrogen-containing compounds, a process that requires energy.

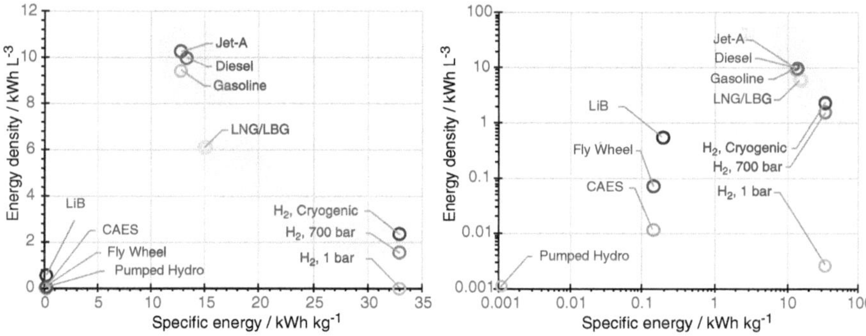

FIGURE 1.1 Energy density and specific energy of relevant energy storage types. On the *left* the values are displayed naturally, whereas on the *right* they are displayed logarithmically. The belt of hydrogen and hydrocarbons relevant in this book is highlighted in *yellow*. (*Gray* in print version)

Molecular hydrogen produced electrolytically from water has an efficiency of approximately 66% (with the present technology), with this fraction of the input energy stored in the produced H_2. However, state-of-the-art electrolyzers that are combined with heat offer a 100% electric energy efficiency, with all of the electrical energy to be stored as molecular hydrogen.

Produced thermochemically (e.g., from either natural gas or coal by steam reforming), energy yield of up to 85% can be expected. The rest of the energy is dissipated in the form of heat. The steam-reforming route from natural gas (essentially methane; CH_4) to hydrogen will also produce on the order of ten times as much CO_2 as H_2 (weight basis), so this will not be a viable route environmentally without large-scale carbon emission management (e.g., CCS).

Biomass is probably the oldest energy storage alternative employed by humans, having been in use since mankind learned to control fire. Burning wood is certainly the oldest chemical process controlled by humans. Biomass may be characterized as having an overall chemical composition in the range from $CH_{1.5}O_{0.7}$ to CH_2O. It is thus evident that hydrogen production from biomass will yield even higher amounts of CO_2 than hydrogen production from natural gas (i.e., CH_4). Still, biomass is considered to be an environmentally viable source of stored energy, for the following reason: as part of the Kyoto agreement, a 100-year time perspective is assumed. For this reason, biomass is considered to be carbon neutral in this time frame because greenhouse emissions from harvested biomass are considered to be assimilated in new biomass grown during such a long time period. Lately, this perspective has been challenged regarding the use of wild forests, like the boreal forest, as a biomass source [5].

It is no longer certain that the Kyoto agreement's 100-years perspective is viable. A shorter time frame is needed in order to assure that a tipping point[a] will not be reached within this century. Furthermore, it has been pointed out that even if biomass is carbon neutral in

[a] The tipping point is a state from which global warming will proceed irreversibly due to positive feedback in the climate system (e.g., the melting of the arctic tundra, with huge emissions of the potent greenhouse gas methane as an unavoidable result).

a 100-year perspective, this does not necessarily mean that it will not cause a net global warming effect over the same period. From when the CO_2 emissions from a given volume of biomass are emitted until all this emitted CO_2 is reassimilated in new biomass, there will be a period where this CO_2 will contribute to increased atmospheric greenhouse gas levels. This time is often referred to as the Carbon Payback Time of biomass, and the corresponding contribution to increased temporal greenhouse gas concentrations in the atmosphere is termed its Carbon Debt.

Recently, significant insights into the temporal CO_2 effect by introduction of the Bio GWP Index has been made [6]. The same group has also pointed out that there are other physical effects besides the carbon emissions that may overshadow the carbon debt considerations for biomass [7]. They stated that the growth of biomass will affect the surface reflectance of the Earth, and a change in biomass growth patterns will alter the portion of incoming high-frequency solar energy that will merely be reflected without transformation to low-frequency heat [7].

Other uncertainty factors related to biomass sustainability include the impacts that growth and harvesting of biomass will have on local and global long-term carbon sequestration (i.e., long-term carbon transfer from biosphere to soil), and the effects that forests have on local precipitation patterns as well as global weather system (e.g., evapotranspiration, and terpene aerosol emissions as rain-forming nuclei). Finally, biomass is used to sustain a wide variety of factors critical to human life and other life forms (e.g., food and materials for shelter). Biomass for energy is thus a limited resource, needing to be applied wisely.

HYDROGEN AND BIOMASS NEXUS

Molecular hydrogen is an energy carrier used for storing energy, and carbon is a hydrogen carrier. These compounds are joined together naturally via photosynthesis, where sunlight is absorbed by various light-harvesting molecules and complexes (e.g., chlorophylls), with the "captured" energy being used to split water into hydrogen and oxygen. In turn, plants capture CO_2 and use it to store hydrogen in hydrocarbon structures (e.g., glucose). In the transition to a carbon-neutral economy, it is essential that society mimic these processes to get access to hydrocarbon chains and capture carbon to utilize them as hydrogen carriers [8,9]. One way to do this is to use renewable electric energy in combination with electrolysis for H_2 production, and then either recombine this hydrogen with CO_2 from air or from biomass.

HYDROGEN BIOMASS DEFICIENCY

In producing biomasses like biogas, large amounts of unfixed concentrated carbon-neutral CO_2 is still present. This lowers the energy content of the gas significantly, and in the liquefaction and biomass to liquid processes this CO_2 is a waste. In biogas, H_2 can be mixed into the gas to increase the energy content. Additionally, in the case of concentrated streams of waste, CO_2, H_2, and CO_2 can be recombined into carbon-neutral liquid hydrocarbons such as diesel and jet-A fuel through various processes (e.g., through a Fischer Trospch reactor). These processes balance the biomass H_2 deficiency to give the highest possible volumetric energy density.

REFERENCES

[1] Lamb JJ, Røkke G, Hohmann-Marriott MF. Chlorophyll fluorescence emission spectroscopy of oxygenic organisms at 77 K. Photosynthetica 2018;56(1): 105–24.
[2] Lamb JJ, Eaton-Rye JJ, Hohmann-Marriott MF. An LED-based fluorometer for chlorophyll quantification in the laboratory and in the field. Photosynth Res 2012;114(1).
[3] Lamb J, Forfang K, Hohmann-Marriott M. A practical solution for 77 K fluorescence measurements based on LED excitation and CCD array detector. PLoS One 2015;10(7.).
[4] Burheim OS. Engineering energy storage. Cambridge, United States, Academic Press; 2017.
[5] Holtsmark B. Harvesting in boreal forests and the biofuel carbon debt. Clim Change 2012;112(2):415–28.
[6] Cherubini F, Peters GP, Berntsen T, Strømman AH, Hertwich E. CO_2 emissions from biomass combustion for bioenergy: atmospheric decay and contribution to global warming. Gcb Bioenergy 2011;3(5):413–26.
[7] Bright RM, Cherubini F, Strømman AH. Climate impacts of bioenergy: Inclusion of carbon cycle and albedo dynamics in life cycle impact assessment. Environ Impact Assess Rev 2012;37:2–11.

[8] Ebbesen SD, Mogensen M. Electrolysis of carbon dioxide in solid oxide electrolysis cells. J Power Sources 2009;193(1):349–58.
[9] Ebbesen SD, Jensen SH, Hauch A, Mogensen MB. High temperature electrolysis in alkaline cells, solid proton conducting cells, and solid oxide cells. Chem Rev 2014;114(21):10697–734.

Chapter | two

Current Use of Bioenergy and Hydrogen

Jacob J. Lamb[a], Bjørn Austbø[b]

[a]*Department of Electronic Systems, Department of Energy and Process Engineering, ENERSENSE, Faculty of Information Technology and Electrical Engineering, NTNU, Trondheim, Norway;* [b]*Department of Energy and Process Engineering, ENERSENSE, Faculty of Engineering, NTNU, Trondheim, Norway*

CHAPTER OUTLINE

ABSTRACT

Fossil fuels have been the cornerstone of our energy infrastructure for a number of centuries now. In recent decades, there has been an emergence of bioenergy and hydrogen as a modern energy source to supply our civilization. Despite this, fossil fuels are still the main energy source for our society, slowing the adaptation of hydrogen and biomass technologies. In this chapter, the history and current utilization of hydrogen and biomass energy are discussed to identify the adaptation of these energy sources.

Hydrogen, Biomass and Bioenergy. http://dx.doi.org/10.1016/B978-0-08-102629-8.00002-5

INTRODUCTION

Fossil-based sources turned into the primary supply of energy for transportation in the 20th century. This has continued into the 21st century with the majority of transport vehicles running on fossil-fuel derivatives [1]. The ongoing utilization of these energy sources to fulfill the global energy demand is dangerous as it increases the concentrations of CO_2 in the atmosphere, provoking a climate change [2].

Moreover, fossil fuels are a finite resource, which is restricted to specific reserves in fixed locations [3]. Taking environmental, economic, and political considerations into account, there has been a growing interest in biofuels for some decades now. Biofuels are created from naturally occurring biomasses and can be used as substitutes for fossil fuels. Additionally, their combustion results in lower sulfuric and aromatic exhaust emissions than typical fuels while providing similar properties in terms of fuel potency [4].

The term biofuel represents liquid or gaseous fuels that are predominantly created from biomass. Large-scale production of biofuels offers a chance for society to scale back their dependence on fossil utilization. Furthermore, in developed countries, there is a growing trend toward using up-to-date technologies and economic bioenergy conversion processes employing a variety of biofuels, which are becoming economically competitive with fossil fuels [5]. For this reason, alternative transportation fuels like bioethanol, biodiesel, and H_2 can play a critical role within the world's energy future.

The production of H_2 is one of the foremost promising energy technologies known to society. This is because it is not energy that exists in large quantities freely in nature. H_2 may be a "secondary" type of energy that acts as storage energy, much like electricity in a battery, resulting in the form of energy that can be stored and distributed. The bulk of the specialists in energy-related fields believe that H_2 will play a large role as an essential primary energy storage and carrier in the future [6,7].

H_2 is clean fuel with no direct CO_2 emissions that can be straightforwardly employed in fuel cells for the generation of electricity. Additionally, H_2 includes a high-energy yield of 122 kJ/g, which is 2.75 times larger than fossil-based fuels [8]. The utilization of H_2 as a fuel for transportation and storage is receiving much favorable attention as environmentally friendly energy [9] and has been extensively explored for use in combustion engines and electric fuel cell vehicles. However, at traditional temperatures and pressures, H_2 is in the form of a gas, which presents more substantial transportation and storage hurdles than those that exist for liquid fuels [10]. Despite this, developments have shown that H_2 may be combined with chemicals in numerous solid and liquid compounds (e.g., metal hydrides, methane, methanol, and lightweight hydrocarbons) [11].

Retrospective and historical development

HYDROGEN

Thirty years ago, H_2 was known as "a crucial and indispensable element of a decarbonized energy system" to produce secure, efficient, and nonpolluting energy [12]. Today, energy leaders see H_2 as the lowest impacting fuel in the world energy system [13], playing a significant role in societies' low-carbon future that will be straightforward to store and be transported [14–17]. This would allow the adaptation of a safer energy system, reducing fossil-fuel dependence [18], with the flexibility to control across the transport [19], heat [20,21], trade [22], and electricity sectors [23]. Together, these sectors currently account for a large fraction of world greenhouse gas emissions [24].

H_2 is created from significant sources of primary energy through numerous production technologies. Most gas is presently created from nonrenewable fossil fuel sources. Half of the current H_2 production is created using thermochemical methods exploiting fossil fuel as an energy source. Of this, petroleum and natural gases provide the most abundant energy supply for H_2 production, followed by coal [25,26]. In short, fossil fuel–based processes account for 95% of the global H_2 production.

The production of H_2 from fossil fuels causes the co-production of greenhouse emissions, that are known to be causing a global climate change. H_2, created through a variety of renewable primary energy sources like wind, biomass, and solar energy, is ideal for step-by-step substitution of fossil fuels [27]. Biomass and biomass-derived fuels can be utilized to manufacture H_2 sustainably. Furthermore, the exploitation of biomass rather than fossil fuels to supply H_2 reduces the global quantity of CO_2 emissions discharged to the atmosphere. This is because the CO_2 emissions discharged was once absorbed from the atmosphere and fixed by chemical reactions within the growing plants [28].

The current annual worldwide H_2 gas consumption is around 400–500 billion Nm^3 [28]. The present utilization of H_2 has a rate of growth calculable at 5%–10% per annum [25]. However, only a fraction of this H_2 is presently used for energy storage and transport purposes; the majority is used as a chemical feedstock for organic compounds, foods, and other process industries. The global marketplace for H_2 is already larger than US$40 billion per annum [29]; composed of utilization in ammonia production (49%), fossil fuel processing (37%), wood production (8%), and miscellaneous smaller volume uses (6%) [30].

A benefit of H_2 is that it can be used as a transportation fuel, whereas neither nuclear nor renewable energy is used directly. H_2 is a fantastic energy storage and transport fuel when coupled with fuel cells, as H_2 can straightforwardly be converted into electricity anywhere, anytime.

BIOENERGY

Biomass, mainly wood, is the oldest sort of energy utilized by humans. Historically, biomass has been utilized through direct combustion, and this method remains widely employed today. The foremost essential biomass energy sources to date are wood and wood wastes, crops and their waste by-products, municipal solid waste, animal wastes, waste from food processes, and aquatic plants and alga. Energy from biomass can be employed directly within the electrical sector, lumber and wood production plants, and pulp and paper industries. Currently, much analysis has been targeted on clean and environmentally friendly energy from biomass sources to transaction from fossil fuels.

Components

The chemical structure and major organic parts in biomass can straightforwardly be utilized within the processes for manufacturing-required fuels and chemicals [31]. The constituents of lignocellulose are the three polymeric components: cellulose, hemicellulose, and lignin [32]. The lignocellulose structure is rigid and obstinate, meaning that the biochemical conversion into usable sugars is rather demanding [33–37]. Generally, the primary three parts have high molecular weights and contribute a lot of mass [38].

Cellulose comes in varied forms with a significant fraction coming from domestic and industrial wastes [28]. Hemicellulose could be a mixture of various polymerized monosaccharides like aldohexose, mannose, galactose, xylose, arabinose, 4-O-methyl glucuronic acid, and galacturonic acid residues [39]. In hardwood xylan, the backbone chain consists of xylose units that are coupled by β-(1,4)-glycosidic bonds and branched by α-(1,2)-glycosidic bonds with 4-O-methyl glucuronic acid groups [38]. Lignin is an aromatic compound synthesized from phenylpropanoid precursors. The essential chemical phenylpropane units of the polymer (primarily syringyl, guaiacyl, and p-hydroxy phenol) are secured together by a group of linkages to make a matrix. This matrix contains a variety of useful groups, like methoxyl and carbonyl that impart a high polarity to the polymer organic compound [40].

Biomass importance

Bioenergy utilization has gained express interest in recent years because of the continued depletion of fossil fuels [41]. The growing interest in bioenergy is driven by [42]:

1. its contribution to poverty reduction in developing countries;
2. meeting the energy requirements cost-effectively;
3. delivering energy in a variety of forms that society demands (e.g., liquid and gaseous fuels for heat and electricity);
4. its low CO_2 footprint; and,
5. its ability to help to revive otherwise unproductive and degraded lands, increasing diversity, soil fertility, and water retention.

The importance of biomass in the energy transition from fossil fuels will increase as global energy policies and strategies focus heavily on decarbonization.

Contribution of bioenergy to the global energy supply

Bioenergy is considered an ideal choice as a transition energy source between the reduction of fossil dependence and the adaptation of an H_2-based society. Most importantly, it can meet the energy requirements for the near future. Biomass energy resources are considered the world's most abundant energy supplier. It is a natural resource comprising 220 billion dry tons (about 4500 EJ = 4500×10^{18} J) of annual primary production [43], yet its utilization as an energy source for our society is currently limited. Although it is not straightforward to access and utilize biomass, the long-term potential for biomass energy utilization is believed to reach 150–400 EJ/year by 2050 (Exajoules/year = 10^{18}J/year) with the employment of forest and industrial residues, in addition to potentially growing permanent energy crops [44].

Currently, much research has been targeted on renewable and environmentally friendly energy from biomass to move away from fossil fuels. Biomass can be utilized in a variety of secondary energy carriers (e.g., electricity; gaseous, liquid, and solid fuels; and heat), employing a large selection of conversion routes and technologies. The current conversion routes to fuels and electricity are distinguished in thermal, chemical, and organic chemistry conversion routes [45]. These include thermo-chemical conversion technologies like pyrolysis [46], gasification [47], liquefaction [48], and supercritical water gasification [49] for maximizing yields. Biomass thermo-chemical conversion technologies like pyrolysis and gasification are by no means the first necessary choices at present, as combustion of biomass is responsible for over 90% of global bioenergy production [50]. In industrialized countries, the most utilized processes using biomass is the direct combustion of residues and wastes for electricity generation [51].

Present uses of hydrogen and biomass

TRANSPORTATION

H_2 and biomass have varying degrees of suitability between transport modes, reflecting the varied nature of the global transport sector (spanning land, ocean and air, and freight and passengers). Nearly half of the energy demand for transport internationally is in the form of light-weight vehicles. Therefore, it is predicted that the number of personal cars worldwide is anticipated to rise from 1 to 2.5 billion by 2050 [52]. With the current dependence of transport on fossil fuels, this increase would prove to be disastrous to our environment if changes are not implemented.

H_2, biofuel, and electric vehicles (EVs) represent the three main choices for low-carbon transport. H_2 avoids the land-use and air quality impacts of biofuels, and the restricted range and long recharging times related to EVs. However, in terms of maturity, EVs are many years ahead of biofuel and H_2 fuel cell cars because of their lower prices and readily available infrastructure. Plug-in EVs currently account for almost 20% of the latest vehicle sales in Norway [53].

In addition to coping with global climate change, H_2 fuel cell vehicles improve air quality. This is a health priority, as over 1,000,000 (> 1 million) premature deaths occur each year across Europe because of particulate emissions in the air [54]. The direct price of pollution because of illness-induced loss of production, healthcare, crop yield, and injury to buildings is around €24billions each year across Europe with indirect prices calculable to be €330–€940billions each year [55].

HEAT AND INDUSTRY

Heating and lighting account for around 60%–80% of the energy consumption in residential and industrial buildings across Europe. To prevent an environmental catastrophe, the emissions generated from heating must be significantly reduced rapidly and, for the most part, eliminated by 2050. However, the heating sector in many regions (e.g., the United Kingdom) is proving challenging to decarbonize due to a variety of reasons [56]:

1. Heating is the largest energy demand in several temperate countries;
2. The requirements for heating are varied significantly, with no method of heat production capable of meeting all heating demands;
3. The heat demand varies daily and seasonally; and,
4. Heating from fossil fuels provides flexibility.

A particular challenge for low-emission heating in temperate countries is meeting the peak winter heating demand that is significantly higher than the peak electricity demand for non-heating purposes. The sizeable seasonal variation between summer and winter is met by traditional gas heating technologies, as their low per-kW costs, and ease of storage and transport make them ideal. Despite this, their CO_2 levels are a sizeable negative consequence of their usage. To reduce utilization of these fuel sources and improve insulation, residential or shared thermal storage and energy storage devices may scale back the peak requirements; however, the required robust policy advocating these adaptations has not been forthcoming in several countries. Furthermore, although various low-carbon choices like electrification or district heating may meet peak heat demand, large investments in infrastructure would be required.

Natural gas is used extensively globally and is presently an inexpensive, convenient, and comparatively clean energy source compared to coal and oil. Due to this, it is the dominant fuel for heating in several countries. In several areas, biogas is also being used and injected directly into natural gas networks to supply fuel for heating. Furthermore, district heating is widely utilized in Russia and Scandinavia [57], permitting centralized energy utilization that is straightforward, distributing biogas as an alternative.

ELECTRICAL ENERGY NETWORK

Global electricity consumption has doubled since 1990 and represents a significant amount of primary energy demand [58]. Fossil fuels produce a significant fraction of global electricity, emitting fifteen $GtCO_2$ annually, or 40% of the world total [59]. Therefore, electricity generation and storage ought to be speedily decarbonized throughout the 2020s, permitting society to push the decarbonization of transport and heating forward through electrification.

Unlike different sectors, electricity generation has already obtained a variety of low-, zero-, and even negative-carbon alternatives. Wind and solar energy are types of intermittent renewable energy. Their output cannot be controlled or predicted as they depend upon the weather, seasons, and time of day. Balancing supply and demand must be further developed if electricity systems are to decarbonize while maintaining current levels of cost and reliability. Moreover, electricity has a significant constraint that production and supply should balance demand, and system reliability is a principal requirement as outages cause severe economic and social damage. In this context, current H_2 and biofuel technologies can assist with the transition to low-carbon renewable electricity generation with the electrification of heating and transport sectors.

INFRASTRUCTURE

The development of H_2 and biofuel infrastructure is a significant barrier to the widespread uptake of these technologies. There is a perception that an "H_2 economy" would be established at a monumental cost without the utilization of existing energy infrastructure. However, various production and distribution pathways already exist (e.g., reforming and methanation), and embrace many progressive steps that do not need complete infrastructure transformations (e.g., utilization of preexisting natural gas infrastructures). By utilizing the current infrastructure, development of a cost-effective infrastructure evolving from these energies will allow the development and production of a fully decarbonized energy sector that our society can transition toward over time.

REFERENCES

[1] Loppacher LJ, Kerr WA. Can biofuels become a global industry?: government policies and trade constraints. Energy Polit 2005;5:7–27.

[2] Balat M. Bioethanol as a vehicular fuel: a critical review. Energy Sources, Part A 2009;31(14):1242–55.

[3] Demirbas MF. Bio-oils from corn stover via supercritical water liquefaction. ENERGY Educ Sci Technol PART A-ENERGY Sci Res 2009;23(1–2): 97–104.

[4] Demirbas A, Dincer K. Sustainable green diesel: a futuristic view. Energy Sources, Part A 2008;30(13):1233–41.

[5] Demirbas A. The importance of bioethanol and biodiesel from biomass. Energy Sources, Part B 2008;3(2):177–85.

[6] Veziroğlu TN, Şahi S. 21st Century's energy: hydrogen energy system. Energy Convers Manag 2008;49(7):1820–31.

[7] Balat M. Potential importance of hydrogen as a future solution to environmental and transportation problems. Int J Hydrogen Energy 2008;33(15): 4013–29.

[8] Kapdan IK, Kargi F. Bio-hydrogen production from waste materials. Enzyme Microb Technol 2006;38(5):569–82.

[9] Cherry RS. A hydrogen utopia? Int J Hydrogen Energy 2004;29(2):125–9.

[10] Ulusarslan D, Gemici Z, Teke I. Currency of district cooling systems and alternative energy sources. ENERGY Educ Sci Technol PART A-ENERGY Sci Res 2009;23 (1–2):31–53.

[11] Shakya BD, Aye L, Musgrave P. Technical feasibility and financial analysis of hybrid wind–photovoltaic system with hydrogen storage for Cooma. Int J Hydrogen Energy 2005;30(1):9–20.

[12] NREL. The green hydrogen report [Online]. Available from: https://www.hydrogen.energy.gov/pdfs/greenhyd.pdf, 1995.

[13] World Energy Issues Monitor, Global and regional perspectives. Available from: https://www.worldenergy.org/assets/downloads/3.-World-Energy-Issues-Monitor-2019-Interactive-Executive-Summary.pdf, 2019.

[14] Brandon NP, Kurban Z. Clean energy and the hydrogen economy. Phil. Trans. R. Soc. A 2017;375:20160400. doi:10.1098/rsta.2016.0400.
[15] Hanley ES, Deane JP, Gallachóir BPÓ. The role of hydrogen in low carbon energy futures–a review of existing perspectives. Renew Sustain Energy Rev 2018;82: 3027–45.
[16] Pudukudy M, Yaakob Z, Mohammad M, Narayanan B, Sopian K. Renewable hydrogen economy in Asia–opportunities and challenges: an overview. Renew Sustain Energy Rev 2014;30:743–57.
[17] Abbasi T, Abbasi SA. 'Renewable'hydrogen: prospects and challenges. Renew Sustain Energy Rev 2011;15(6):3034–40.
[18] Dunn S. Hydrogen futures: toward a sustainable energy system. Int J Hydrogen Energy 2002;27(3):235–64.
[19] Tollefson J. Hydrogen vehicles: fuel of the future? Nat News 2010;464(7293): 1262–4.
[20] Dodds PE, Demoullin S. Conversion of the UK gas system to transport hydrogen. Int J Hydrogen Energy 2013;38(18):7189–200.
[21] Dodds PE, Staffell I, Hawkes AD, Li F, Grünewald P, McDowall W, et al. Hydrogen and fuel cell technologies for heating: a review. Int J Hydrogen Energy 2015;40(5):2065–83.
[22] Napp TA, Gambhir A, Hills TP, Florin N, Fennell PS. A review of the technologies, economics and policy instruments for decarbonising energy-intensive manufacturing industries. Renew Sustain Energy Rev 2014;30: 616–40.
[23] Ball M, Weeda M. The hydrogen economy–vision or reality? Int J Hydrogen Energy 2015;40(25):7903–19.
[24] World resources institute. CAIT Climate data explorer [Internet]. Available from: https://www.wri.org/our-work/project/cait-climate-data-explorer, 2019.
[25] Mohan SV, Bhaskar YV, Sarma PN. Biohydrogen production from chemical wastewater treatment in biofilm configured reactor operated in periodic discontinuous batch mode by selectively enriched anaerobic mixed consortia. Water Res 2007;41(12):2652–64.
[26] Das D, Khanna N, Veziroğlu NT. Recent developments in biological hydrogen production processes. Chem Ind Chem Eng Quarterly/CICEQ 2008;14(2): 57–67.
[27] Yolcular S. Hydrogen production for energy use in European Union countries and Turkey. Energy Sources, Part A 2009;31(15):1329–37.
[28] Balat H, Kırtay E. Hydrogen from biomass–present scenario and future prospects. Int J Hydrogen Energy 2010;35(14):7416–26.
[29] Kraus T. Hydrogen Fuel—an economically viable future for the transportation industry? Duke J Econ 2007;19:39.
[30] Konieczny A, Mondal K, Wiltowski T, Dydo P. Catalyst development for thermocatalytic decomposition of methane to hydrogen. Int J Hydrogen Energy 2008;33(1): 264–72.

[31] Balat M. Hydrogen in fueled systems and the significance of hydrogen in vehicular transportation. Energy Source 2007;2(1):49–61.
[32] Yaman S. Pyrolysis of biomass to produce fuels and chemical feedstocks. Energy Convers Manag 2004;45(5):651–71.
[33] Horn SJ, Vaaje-Kolstad G, Westereng B, Eijsink V. Novel enzymes for the degradation of cellulose. Biotechnol Biofuels 2012;5(1):45.
[34] Quinlan RJ, Sweeney MD, Lo LL, Otten H, Poulsen JC, Johansen KS, et al. Insights into the oxidative degradation of cellulose by a copper metalloenzyme that exploits biomass components. Proc Natl Acad Sci 2011;108(37): 15079–84.
[35] Vaaje-Kolstad G, Westereng B, Horn SJ, Liu Z, Zhai H, Sørlie M, et al. An oxidative enzyme boosting the enzymatic conversion of recalcitrant polysaccharides. Science 2010;330(6001):219–22.
[36] Vivekanand V, Olsen EF, Eijsink VGH, Horn SJ. Effect of different steam explosion conditions on methane potential and enzymatic saccharification of birch. Bioresour Technol 2013;127:343–9.
[37] Sarker S, Lamb JJ, Hjelme DR, Lien KM. A review of the role of critical parameters in the design and operation of biogas production plants. Appl Sci 2019;9(9):1915.
[38] Demirbas A. Pyrolysis of biomass for fuels and chemicals. Energy Source 2009;31(12):1028–37.
[39] Mohan D, Pittman CU, Steele PH. Pyrolysis of wood/biomass for bio-oil: a critical review. Energy Fuels 2006;20(3):848–89.
[40] Feldman D, Banu D, Natansohn A, Wang J. Structure–properties relations of thermally cured epoxy–lignin polyblends. J Appl Polym Sci 1991;42(6): 1537–50.
[41] Caputo AC, Palumbo M, Pelagagge PM, Scacchia F. Economics of biomass energy utilization in combustion and gasification plants: effects of logistic variables. Biomass Bioener 2005;28(1):35–51.
[42] Karekezi S, Lata K, Coelho ST. Traditional biomass energy: Improving its use and moving to modern energy use. Renew Energy—A Glob Rev Technol Policies Mark 2006;1:231–61.
[43] Jekayinfa SO, Scholz V. Potential availability of energetically usable crop residues in Nigeria. Energy Sources. Part A Recover Util Environ Eff 2009;31(8):687–97.
[44] Charles MB, Ryan R, Oloruntoba R, von der Heidt T, Ryan N. The EU–Africa energy partnership: towards a mutually beneficial renewable transport energy alliance? Energy Policy 2009;37(12):5546–56.
[45] Hoogwijk M, Faaij A, Eickhout B, de Vries B, Turkenburg W. Potential of biomass energy out to 2100, for four IPCC SRES land-use scenarios. Biomass Bioenergy 2005;29(4):225–57.
[46] Caglar A. Valorization of tea wastes by pyrolysis. ENERGY Educ Sci Technol PART A-ENERGY Sci Res 2009;23(1–2):135–44.
[47] Bhuiyan MNA, Ota M, Murakami K, Yoshida H. Pyrolysis kinetics of newspaper and its gasification. Energy Sources, Part A Recover Util Environ Eff 2009;32(2): 108–18.

[48] Demirbas A. Liquefaction of biomass using glycerol. Energy Source 2008;30(12): 1120–6.
[49] Şener Ü, Genel Y, Saka C, Kiliçel F, Küçük MM. Supercritical fluid extraction of cotton stalks. Energy Source 2009;32(1):20–5.
[50] Demirbas A. Biofuels from agricultural biomass. Energy Source 2009;31(17): 1573–82.
[51] Bakis R. Alternative electricity generation opportunities. Energy Source 2007;30(2): 141–8.
[52] US Energy information administration, International energy outlook 2017. Available from: https://www.eia.gov/outlooks/ieo/pdf/0484(2017).pdf, 2017.
[53] Norsk elbilforening. Elbilsalget: Ned i fjor – venter ny vekst i år. 2017.
[54] Walton H, Dajnak D, Beevers S, Williams M, Watkiss P, Hunt A. Understanding the health impacts of air pollution in London. London Kings Coll London, Transp London Gt London Auth; 2015.
[55] European commission. A clean air program for Europe. 2013.
[56] Chaudry M, Abeysekera M, Hosseini SHR, Jenkins N, Wu J. Uncertainties in decarbonising heat in the UK. Energy Policy 2015;87:623–40.
[57] Hanna RF, Gross R, Parrish B. Best practice in heat decarbonisation policy: a review of the international experience of policies to promote the uptake of low-carbon heat supply. 2016.
[58] BP. BP statistical review of world energy. 2017.
[59] International energy agency. World energy outlook. 2016.

Chapter | three

Traditional Routes for Hydrogen Production and Carbon Conversion

Jacob J. Lamb[a], Magne Hillestad[b], Erling Rytter[b], Robert Bock[c], Anna S.R. Nordgård[d], Kristian M. Lien[c], Odne S. Burheim[e], Bruno G. Pollet[e]

[a]*Department of Electronic Systems, Department of Energy and Process Engineering, ENERSENSE, Faculty of Information Technology and Electrical Engineering, NTNU, Trondheim, Norway;* [b]*Department of Chemical Engineering, ENERSENSE, Faculty of Natural Sciences, NTNU, Trondheim, Norway;* [c]*Department of Energy and Process Engineering, ENERSENSE, Faculty of Engineering, NTNU, Trondheim, Norway;* [d]*Department of Biotechnology and Food Science, Faculty of Natural Sciences, NTNU, Trondheim, Norway;* [e]*Department of Energy and Process Engineering, ENERSENSE & NTNU Team Hydrogen, Faculty of Engineering, NTNU, Trondheim, Norway*

CHAPTER OUTLINE

Hydrogen, Biomass and Bioenergy. http://dx.doi.org/10.1016/B978-0-08-102629-8.00003-7

ABSTRACT

There are many routes to produce hydrogen. Traditionally, hydrogen can be produced through electro-, bio-, and thermo-chemical routes. Biochemical routes involve routes such as biophotolysis, fermentation, or through biological metabolic pathways. Electrochemical production methods of hydrogen mainly involve water electrolysis (the splitting of water into molecular hydrogen and oxygen). The thermochemical production of hydrogen is achieved though reforming, pyrolysis, and gasification reactions of biomass and fossil fuels (carbonaceous fuels). This chapter gives an overview of the methods of hydrogen production through these main routes, as well as providing novel methods that are not yet widely used.

THE BIOCHEMICAL ROUTE

Microorganisms catalyze the biochemical route of H_2 production as opposed to other methods that require chemical or electrochemical counterparts. They have generally suited production methodologies for decentralized, small-scale installations that are close to sources of biomass or biowaste. These processes can be carried out by either anaerobic bacteria, algae, or a combination of both, and the characteristics can differ significantly between different microorganisms and substrates used. The traditional biological H_2 production processes can be characterized into the following groups, as shown in Table 3.1.

Use of microorganisms for biophotolysis of water

Oxygenic photosynthesis is a light-driven process in which H_2O is oxidized to O_2, and electrons released are used for nicotinamide adenine dinucleotide phosphate (NADP) and ferredoxin reduction. Adenosine triphosphate (ATP) is then produced from the protons (H+) released in water oxidation, and those that are transported across the photosynthetic membranes by proton pumps. The reducing equivalents NADPH (reduced form of NADP) and ferredoxin can be used to facilitate the

TABLE 3.1 Traditional Biological Hydrogen Production Processes

Process	Description
Direct biophotolysis	The biological process of using solar energy through the photosynthetic machinery directly to split water (H_2O), thus evolving H_2 and O_2
Indirect biophotolysis	The biological production of carbohydrates through photosynthesis followed by conversion into H_2 through hydrogenases in anaerobic conditions
Photofermentation	The biological conversion of organic substrates into H_2 via light-driven fermentation
Dark fermentation	The biological conversion of organic substrates into H_2 via fermentation in the absence of light
Microbial electrolysis cell	Microbial-based fuel cells that have an external electrical current input to boost the production of H_2 from organic materials

production of H_2 by hydrogenases (present in many algae and cyanobacteria).

In a standard photosynthetic system, there is a requirement of four photons to remove a pair of electrons from H_2O. A further two photons are required to then drive the electron transfer through photosystem II (PSII), bringing the electrons to a redox potential (E^o) of 0 V, as opposed to the + 0.82 V redox potential of water (where they originated). Another two photons can then be utilized to transfer the electrons from PSII to photosystem I (PSI). This increases the reduction potential of the intermediates on the acceptor side (PSI side) enough to allow the reduction of protons to H_2 (about –0.4 V) [1–3].

The overall net equation of photosynthetically produced H_2 via hydrogenases can be observed as the following:

$$2H_2O \rightarrow 2H_2 + O_2 \tag{3.1}$$

This is facilitated by The Earth's energy input from The Sun in the form of photosynthetically active light (PAR; 400–700 nm electromagnetic radiation) [4]. Because this pathway does not involve any emission of carbon, this is a desirable method of H_2 production. Therefore, the connection of photosynthesis to The Sun's energy input (can be assumed to be the most significant renewable and sustainable source of energy for our planet) links H_2 to a truly renewable and sustainable source. Despite this idea of production of H_2 through oxygenic photosynthesis being highly attractive, it is practically and experimentally challenging to achieve.

In principle, the absorption of a photon by PSII is sufficient to drive H_2 production, because this absorbed photon leads to the formation of pheophytin (midpoint redox potential of –0.6 V). Practically, however, reduced pheophytin is very unstable with a lifetime in the range of nanoseconds, and the subsequent transfer of electrons to hydrogenase would need to be enhanced by the engineering of the proteins that facilitate the transfer [5]. This single-complex catalyst, although only somewhat theoretical, would allow for a significant amount of H_2 to be produced from photosynthesis, due to the need for only one photosystem.

Despite which route the electrons take from the water-splitting complex in PSII to hydrogenases in oxygenic photosynthetic organisms, the hydrogenases themselves are very sensitive to even low levels of oxygen [6,7]. In a biological and evolutionary context, this makes sense, because it is of little value for the organism to produce a readily diffusive gas from its valuable reducing agents, rather than using them for respiration to produce energy equivalents like ATP. Since O_2 is an inevitable product of H_2O splitting, the tolerance of hydrogenases for O_2 needs to be drastically improved if oxygenic photosynthesis is to become practical. Possible methods to achieve this could be by either performing genetic modifications of the native hydrogenases to improve their oxygen sensitivity [8] or by the insertion of non-native hydrogenases from other organisms that have oxygen-tolerance [9,10].

Hydrogenases themselves come in two common forms of FeFe hydrogenases (these types occur in both Bacteria and Eukarya), and NiFe hydrogenase (these occur in Archaea and Bacteria) [11]. The sensitivity of FeFe hydrogenases to oxygen is exceptionally high, and they tend to denature in minimal concentrations of oxygen; however, the sensitivity to oxygen is reduced in NiFe hydrogenases, in which these might remain active for several minutes despite being in the presence of oxygen. The NiFe hydrogenases' interaction with O_2 is also reversible [12]; therefore, protein re-synthesis would not be required after O_2 exposure. Although FeFe hydrogenases have the potential to catalyze H_2 production at a higher rate, the O_2 tolerance exhibited by NiFe hydrogenases is more favorable for protein engineering into oxygenic phototrophs. The non-photosynthetic bacteria *Desulfovibrio fructosovorens* was shown to have an increased tolerance to O_2 with the modification of only one amino acid within their hydrogenase gene [13].

The energy equivalent NADP(H) or reduced ferredoxin are usually the sources of electrons for hydrogenases. Despite this, the oxidation of NADPH to form H_2 is not energetically favorable due to the redox midpoint of H_2/H^+ being more negative than that of NADPH/NADP. Therefore, for this reaction to occur, the NADP pool needs to be close to 100% in its reduced form (NADPH), or the concentration of H_2 needs to be very low [14]. As an electron donor, ferredoxin is more suitable for hydrogenases, as the redox midpoint of ferredoxin is more negative than that of H_2/H^+. With this in mind, it seems clear that a hydrogenase with a high level of O_2 tolerance and dependent on ferredoxin is desirable.

A stronger enhancement to the performance of hydrogenase may be a direct coupling of the hydrogenase to PSII. This has been attempted but resulted in reduced electron-transfer rates between PSII and the hydrogenase [15,16]. Better structure information regarding PSII has since been resolved, and this may facilitate improvements of direct coupling of the hydrogenase with PSII [5]. Experiments with Pt (platinum) nanoparticles as the catalyst for H_2 production displayed better rates of production when tethered directly to PSI by a short-carbon chain in vitro [17]. Despite the appeal of this method, it cannot be achieved when integrated into a self-renewing and self-repairing biological system; however, it does display how efficient the electron flow from PSI to H_2 can be.

The use of biological systems for H_2 production powered by solar energy is very appealing because the substrate (water) is abundant, and we have high levels of solar irradiance incident with The Earth's surface. Despite this, several decades of research have not resulted in much of an improvement in this form of biological production of H_2, and many technical barriers seem to be persistent [18–22].

Fermentative production of biohydrogen

By using protons as an electron sink during dark fermentation, bacteria can produce H_2 from organic substrates. The production via fermentation uses the pyruvate-ferredoxin hydrogenase or the pyruvate-formate lyase. Due to the source of organic compounds being biomass, this form of H_2 production can be considered renewable, since all biomass inevitably originates from photosynthesis.

The release of two electron-equivalents by the oxidation of pyruvate to acetyl-CoA + CO_2 and of formate to CO_2 during fermentation is what drives to the formation of H_2 [23,24]. The fermentation to produce H_2 is performed by either strict anaerobes (*Clostridium, Ethanoligenens,* or *Desulfovibrio*) or facultative anaerobes (*Enterobacter, Citrobacter, Klebsiella, Escherichia coli,* or *Bacillus*) [24–27]. In strict anaerobes, electrons are obtained from pyruvate oxidation, transferred to ferredoxin and then to the H_2 catalyzing hydrogenase [5,24]. Formate oxidation is the primary source of electrons for facultative anaerobes when producing H_2, a reaction catalyzed by the enzyme formate-hydrogen lyase [25,27,28].

The majority of electrons produced in the fermentation of organic substrates do not end up reducing H^+ to H_2; they instead are routed toward soluble organic products. Typically, only about 17% of the donor electrons end up facilitating H_2 evolution [23], leading to 2 moles of H_2 per mole of glucose excepted as the typical maximum H_2 yield in mixed-culture dark fermentation [29]. In mixed-culture fermentation, the microorganism systems are classified as either acetate/butyrate (*Clostridium* species [5,30–32]) or acetate/ethanol (*Ethanoligenens* species [26]) groups based on the dominant organic products. The pH of the reaction seems to be the determining factor in what group is more dominant (acetate/butyrate = pH 5–6; acetate/ethanol = pH 4.5 [5,26,29–32]). Nevertheless, the pyruvate-ferredoxin pathway is essential for forming H_2 in both cases.

Compared to other forms of biological H_2 production, the rate at which dark fermentation can produce H_2 is far superior and can even be orders of magnitude higher [29,33,34]. Due to the retention of H_2-producing bacteria in dark fermentation systems, the biomass concentrations can also be as high as 25 g of volatile suspended solids/L [35]. This allows the H_2 production rate to reach up to 7.9 L H_2/L/h, at 35.8°C, 1 atm, and a hydraulic retention time of only 0.5 h [35]. This dramatically reduces the required size of the fermenting reactor by 40–80 times to acquire the same energy output when compared to photosynthetic and microbial electrolysis techniques [36,37].

Despite the production rates of dark fermentation to produce H_2, the yield is low. If 100% conversion of glucose were possible, production of 12 moles of H_2 per mole of glucose would be obtained. However, based upon known fermentation dynamics, the theoretical maximum yield of H_2 is more in the range of 4 moles of H_2 per mole of glucose (when the only electron sinks are acetate and H_2) [24]. Realistically, many other organic electron sinks are present and generated in significant amounts, lowering the actual yield of H_2 further to approximately 2 moles of H_2 per mole of glucose during mesophilic, acidic fermentation [23,29–32].

Not only does the production of these organic products reduce the H_2 yield, but they can also have negative consequences on the system as a whole due to their high biochemical oxygen demand (BOD). This means the effluent requires treatment before being released into the natural environment. These organic products can also have strong odors (e.g., butyrate), or can reduce sulfates and become highly toxic (e.g., hydrogen sulfide) [29]. Therefore, efficient methods of capturing the electron equivalents and the potential energy of soluble organic products are essential for dark fermentation production of H_2 to become viable.

The H_2 that is formed by dark fermentation can be readily converted into undesired products before it has been harvested. This risk is highest in the presence of H_2-oxidizing methanogens and homo-acetogenins, which produce methane (CH_4) and acetate, respectively, from the oxidation of H_2 and reduction of CO_2. The inhibition of the growth of methanogens and homo-acetogens can be achieved at a low pH [38,39]. This may not entirely inhibit methanogens and homo-acetogens as some methanogens are known to be acid-tolerant [40]. Thermal pretreatment (90–100°C for 15–30 mins) of the dark fermentation inoculum represents another method for reduction of methanogens and homo-acetogens, as it actively selects for H_2 producers, which are spore-forming organisms and, therefore, have a higher tolerance to high temperatures [29,31]. This method is not so useful when the input substrate is rich in H_2-consuming microorganisms; therefore, the best method for reducing the oxidation of H_2 is by increasing the harvesting rate to match the H_2 production rate.

The Wood-Ljungdahl pathway

Acetate may be biologically produced via two pathways—β-oxidation or the Wood-Ljungdahl (WL) pathway [41,42]. In β-oxidation, fatty acids are degraded. In the process, one molecule of acetate and one proton is produced per round of β-oxidation until only acetate or propionate remain. The remaining propionate is degraded to acetate by coupling two molecules of propionate to a C6 molecule with the subsequent production of three molecules of acetate [43]. The WL pathway enables acetogenic bacteria to use H_2 as an electron donor and CO_2 as an electron acceptor and as building blocks for biosynthesis. The pathway may be viewed as a series of reactions where two molecules of CO_2 are reduced and four molecules of H_2 are oxidized to produce acetyl-CoA and eventually acetate. The WL pathway is also referred to as the reductive acetyl-coenzyme A pathway. The WL pathway is utilized in its reductive direction by acetogenic bacteria, in its oxidative direction (oxidation of acetate to H_2 and CO_2) by syntrophic acetate-oxidizing bacteria (SAOB) and certain sulfate reducers, and even in part by aceticlastic methanogenic archaea [41,42]. Acetogenic bacteria may be divided into two groups based upon energy conservation [41]; Na^+-dependent with *Acetobacterium woodii*, *Thermoanaerobacter kivui*, and *Ruminococcus productus* as examples, and H^+-dependent with *Moorella* spp. as examples.

THE THERMOCHEMICAL ROUTE

The thermochemical processes of H_2 production involve thermally assisted chemical reactions that release H_2 and H_2-rich gases. Bio-mass, nuclear, and fossil fuels can be used to produce H_2 via this process, and depending on the sources, the thermochemical routes can widely differ [44]. The traditional thermochemical H_2 production processes from various sources broadly include reforming [45], pyrolysis [46,47], gasification [47], water splitting [48,49], supercritical water extraction [50], and catalytic decomposition [51].

Reforming

Reforming refers to the conversion of carbonaceous fuels using various techniques. When hydrocarbon fuel reacts with steam or O_2 (or a combination of both) at high temperatures, H_2 is produced, and the reactions are called reforming reactions [52]. When only steam is used, the reaction is endothermic, and the process is coined steam reforming [53]. The partial oxidation reaction uses O_2 for the fuel conversion and is known as an exothermic reforming reaction [54], and the combination of both steam reforming and partial oxidation is known as an auto-thermal reforming reaction [55].

Hydrocarbon fuel natural gas is characteristically an excellent candidate for reforming to H_2, as this source contains a high hydrogen-to-carbon ratio, which minimizes the formation of CO_2 and is easy to handle for usage [56]. Coal and oil can also be reformed to H_2. However, these feedstocks, unlike natural gas, have a meager hydrogen-to-carbon ratio, and thereby need a large portion of the produced H_2 to be supplied from steam [52]. Among the renewable sources, biomass is the only fuel made of carbon and hydrogen; therefore, this feedstock has excellent potential for H_2 production via steam-reforming conversion.

A typical reforming plant consists of a series of components (e.g., reforming and cleaning units, desulfurization units, compressors, pumps, coolers, heat exchangers) [52]. The plant design is primarily influenced by the operating parameters like temperature, catalyst, steam-to-hydrocarbon ratio, pressure, and the chemical properties of the feedstock. For example, subject to the fuel sulfur content, a desulfurization unit, which reduces sulfur to prevent catalyst poisoning, preceding to the reformer reactor may be added to the plant. Also, depending on the type of fuel, the steam flow required for reforming may be affected, which further influences the size of the reformer and the type of cleaning units. The various reforming methods, along with approaches related to renewable sources, are briefly discussed further.

STEAM REFORMING

Steam reforming (SR) occurs at high temperature with the presence of a catalyst converting hydrocarbons and steam to H_2 and other gases. Since steam reforming is reversible and strongly endothermic (according to the

Le Chatelier principle [57]), it must be carried out at high temperatures, high steam to fuel ratio, and low pressure to maximize yield. The main steps in steam reforming include synthesis gas generation, the water gas shift reaction (WGS), and gas purification, as shown in the following reactions [58,59]:

$$C_nH_m + nH_2O \leftrightarrow nCO + \left(n + \frac{1}{2}m\right)H_2 \left(\Delta H^{\circ}_{298} = -1109\ kJ/mol\ f\ or\ nC_7H_{16}\right) \quad (3.2)$$

$$CO + H_2O \leftrightarrow CO_2 + H_2 \left(\Delta H^{\circ}_{298} = -41\ \text{kJ/mol}\right) \quad (3.3)$$

$$CO + 3H_2 \leftrightarrow CH_4 + H_2O \ \left(\Delta H^{\circ}_{298} = -206\ \text{kJ/mol}\right) \quad (3.4)$$

$$CO_2 + 4H_2 \leftrightarrow CH_4 + 2H_2O\left(\Delta H^{\circ}_{298} = -165\ \text{kJ/mol}\right) \quad (3.5)$$

A wide range of fuels can be used for steam reforming, the choice of which depends on the desired volume of production and the cost of the feedstocks. For a large-scale industrial operation, steam reforming of natural gas [or steam methane reforming (SMR)], is generally preferred, as a high conversion efficiency in the order of 75%–85% through this route can be achieved [52]. However, since natural gas is not a readily available fuel everywhere, and there is an increasing effort of maximizing renewable energy production, an intermediate step converting biomass and its derivatives to natural gas before steam reforming is becoming a promising option [46]. Heavy oil fractions (due to their high availability), are also becoming increasingly considered. Other fuels, such as various combinations of the light hydrocarbonethane, propane, butane, and light and heavy naphtha, have long been used for steam reforming applications.

The catalysts used for reforming widely differ depending on the feedstock type; however, a nickel-based catalyst is the most common for SMR. The performance and cost apparently are the most critical considerations in catalyst selection, and there can be a combination of a range of noble (e.g., Rh, Ru, Pt, Pd, and Ir), transition (e.g., Ni, Co, Cu, and Fe), and oxide support metals (e.g., Ni/MgO, Ru/Mg(Al)O, and Ni.Al_2O_3) suitable for catalyst composition [60,61]. A catalyst consisting of noble metals has proved to be highly active and stable toward steam reforming reactions [62]. However, their high cost drives attention toward low-cost materials instead (e.g., nickel-based catalysts), which gives a similar activity as the former. Apart from metallic catalysts, non-metallic catalysts were also researched; however, due to their low activity, their commercial success is not yet confirmed [63].

The reformer employed for SR can be of various types that are typically categorized as either a tubular reformer, an adiabatic prereformer, or a membrane reformer [52,58]. A reformer is essentially a heat exchanger that produces a high-outlet temperature of around 800–950°C (satisfying the high fuel utilization to H_2), via heating media sourced either at the top, bottom, or side wall of the tubes [64]. Heat to the process is usually transported using convection, radiation, or internal combustion, where radiant heat transfer is the most predominant in industrial operations [58]. A tubular reformer is a radiant heat transfer device designed to produce a high volume of H_2 with an hourly capacity exceeding 300,000 Nm^3, where the heat is passed over a nickel-based catalyst. A prereformer is an extension of a tubular reformer and primarily aimed to reduce the heat requirement in the main tubular reactor by increasing the preheat via conversion of heavy hydrocarbons, lowering the operational cost. A promising alternative of the tubular reformer is the membrane reformer [65,66]. With this configuration a selective membrane (mainly made of palladium) can be integrated either directly to the reaction environment or downstream to separate H_2 by absorption and atomic dissociation at substantially lower temperatures (450–550°C), with the same conversion efficiency (90%–95%) as achieved in a conventional SMR [66]. Research toward further developments of membrane reformers is ongoing, and some of the approaches such as fluidized bed and circulating fluidized bed membrane reactors are already existing and established [67,68].

The fuel first goes into the reformer, where due to high temperatures (around 800–950°C), pressure up to 35 bar, a steam-to-carbon ratio of ~3.5, and, in the presence of a catalyst, production of a gas consisting mostly of CO and H_2 is produced (eq. 3.2). The operating conditions selected in the reformer are adjusted in such a way that catalyst cooking is avoided, and the desired gas yield is achieved. After the reformer, the gas mixture passes through a WGS unit, where an exothermic shift reaction converts the CO into additional H_2 (eq. 3.3) and exchanges the excess heat to the heat recovery unit for preheating of the steam. The shift conversion can take place in different steps in temperatures between 230 and 370°C. As CO_2 is one of the by-products in WGS, the produced gas is fed to the gas purification unit where CO_2 is either absorbed by an amine or separated through pressure swing adsorption (PSA) or a membrane resulting in a pure H_2 stream (near 100% [69]). Finally, in order to convert the residual CO and CO_2 in the gas stream back to methane, exothermic methanation (Equations 3.4 and 3.5) is employed, which occurs at temperatures of around 370–425°C. The energy requirement of a conventional SMR was measured and found as 63 kJ/mol H_2 [52]. This corresponds roughly to 30%–35% of the energy total of the natural gas input into the process.

PARTIAL OXIDATION REFORMING

Methane might also be regenerated into an H_2-rich gas via partial oxidation reforming (POR), which can be either catalyzed or non-catalyzed. POR could also be expressed in several ways. If CH_4 is converted to H_2 and CO_2, the reaction is exoergic.

$$CH_4 + O_2 \rightarrow 2CO_2 + 2H_2 \left(\Delta H^0 = -320 \text{ kJ/mol}\right) \tag{3.6}$$

Several metals are used as catalysts in partial oxidation reforming; however, nickel (Ni) is usually the most popular and economical [70]. In the industry, nickel catalyst is supported on a solid compound. The cost of the catalyst needed drives the selection of the support. Moreover, because stability within the operation is desired, numerous types of aluminum oxide are used as support. The most stable type of aluminum oxide used here is corundum (or α-alumina) and serves as a perfect support for reforming applications where there is not a requirement for the high-specific activity of the catalyst.

For applications requiring high specific activity, the catalyst is prepared using co-precipitation techniques. Both aluminium oxide and nickel are co-precipitated by addition of alkali to a solution of their salts as hydroxycarbonates. This is later separated by washing, filtering, and drying to decompose the precursors. However, this technique results in the less thermodynamically stable form, γ-alumina. Under high-partial pressures of steam at elevated temperatures, this leads to hydrothermal sintering of the γ-alumina, making significant changes to the microstructure of the catalyst, resulting in its eventual degradation. Despite this, the POR method has evolved significantly over the years to maximize the activity and rate of the heat transfer within the reactor.

AUTOTHERMAL REFORMING

Autothermal reforming (ATR) is a combination of SR and POR (both reactions manifest themselves in one compact reactor). The ATR reactor vessel contains a combustion zone and a set catalyst bed among a refractory-lined pressure shell. In ATR, H_2 or H_2-enriched gas is injected along with steam to produce CH_4. ATR is generally operated between temperatures of 950–1050°C, a pressure of 30–50 bar, a steam-to-carbon molar ratio (S/C ratio) of 0.5–1.5, and an oxygen-to-carbon molar ratio (O/C ratio) of 0.6–1.0 [71–73].

Pyrolysis

Among the various thermochemical routes aforementioned for H_2 production from biomass, the two-stage pyrolysis-reforming (PR) strategy has been gaining increasing attention recently. It's many benefits relative to the PR process is the separation of the transformation and reforming steps in numerous parts of the process. Therefore, this method permits for independent temperature amendments within the separate transformation and reforming steps [74]. Moreover, compared to other processes, the direct contact between the feedstock and its impurities with the reforming catalyst is avoided [75].

Additionally, the reforming step is performed at a lower temperature, so catalyst deactivation by sintering is avoided [76,77], particularly once nickel is employed as a catalyst. Moreover, the process does not specialize in H_2 production, as syngas is the main product. The quality and enrichment of the syngas depend on the gasifying agent used, and additional processes are required to boost the H_2 production. Moreover, extremely active steam reforming catalysts allow the complete conversion of biomass-derived volatiles, with the gas product being released from tars, and this may be an advantage over typical methods [78].

Regarding bio-oil reforming, volatiles from biomass are valorized in-line, and so condensation and volatilization of the intermediate compounds from biomass transformation do not seem to be required. This avoids the handling of these products, mass losses, and repolymerization throughout the raw bio-oil volatilization step. Moreover, the most significant advantage of the PR method is its ease of scaling up. Although PR is extensively studied within the literature, there are significant operational issues observed by some authors in these processes [79,80]. Therefore, product distribution, the sort of the reactor for each transformation, and reforming steps and their configuration play an essential role for the method.

Biomass steam gasification

Biomass steam gasification (BSG) has been widely studied throughout recent decades. This is because gases can be directly used as intermediate products for the large-scale production of fuels and chemicals [81–84]. The method characteristics entail establishing plants within regions where biomass industries are present; as a result, prices for the transportation of the material would be excessive [85]. BSG usually takes place at temperatures between 700 and 1200°C, using air, oxygen, steam, or mixtures of these as the gasifying agent. This ends up as a gas product primarily as H_2, CO, CO_2, CH_4, and other hydrocarbons. The utilzation of steam enhances H_2 formation and produces a valuable heating gas with no nitrogen. Although BSG is extremely energy demanding, it avoids the requirement for expensive gas separation methods [86].

The main steps in BSG are pyrolysis, drying, char reactions, and specific phase change reactions (i.e., reforming, cracking, and WGS reactions). In the drying method, the evaporation of the moisture within the biomass occurs. This is typically a quick step and happens at relatively low temperatures, with its influence on the other parts of the process being restricted. Within the phase-changing methods, biomass is thermally degraded to yield gases, condensable products, and char. The liquid product or primary tar produced in biomass reforming could be an advanced mixture of oxygenates created from alcohols, furans, ketones, saccharides, acids, and phenols [87–89]. However, these compounds are generally unstable, and during reforming conditions, they are quickly cracked, changed, or evolved toward stable aromatic structures like secondary and tertiary tars [90,91]. The tar contained within the syngas could be a significant problem for its valorization (providing it is polymerized or condensed below 300°C), potentially causing fouling or corrosion, leading to blockage of pipes, heat exchangers, and particle filters [92,93]. Therefore, the development of tar cracking processes is one of the most significant challenges within this field [94]. Moreover, the total elimination of the main tertiary tars needs temperatures of around 1250°C and residence times above 0.5 s [91].

The char produced within the reformation steps is vaporized by steam and Boudouard reactions; however, the speed of each reaction is low. Therefore, the char reaction represents the limiting step in the process and adequate reactor design with an acceptable experimental duration to achieve a high-char conversion [95–97].

The BSG reaction products are as follows—a gas fraction (usually composed of 30%–50% v/v of H_2, 25%–40% v/v of CO, 8%–20% v/v of CO_2, and 6%–15% v/v of CH_4); a heavy fraction (tar; that consists of a mixture of aromatic hydrocarbons); and a solid fraction (derived from the biomass that has not been entirely vaporized). The yields and properties of these products are influenced by a variety of things, with the leading influencers being reactor configuration (which determines the contact mode, and thus mass and warmth transfer rates), initial biomass characteristics, operational conditions (such as temperature and steam to biomass ratio), and also the use of catalysts [84,86,98].

Supercritical water gasification

Supercritical water gasification (SWG) is different as compared to SR processes. This is because no drying pretreatment is needed for the wet content of the biomass (typically >30% w/w; like food wastes [99], algae [100], or waste material sludge) [101]. SWG takes place at temperatures above 374°C and pressures above 22.1 MPa [50,102], with H_2 production increasing at temperatures above 600°C (lower temperatures are reported to increase CH_4 concentration) [50,102].

However, the SWG method presents serious drawbacks for large-scale feasibility. Pumping of the feedstock is a technological challenge, as biomass must be resuspended into a pumpable liquid, limiting the dry biomass content within the suspension to 20% w/w. Additionally, plugging by char, tar, or base-forming catalysts might occur in long experimental runs, and the severe conditions might cause material corrosion issues [103]. Finally, the high-energy consumption of this method is one of the crucial drawbacks, as water needs to be maintained at critical conditions.

These drawbacks have hindered the quantifiability of the SWG method, resulting in the use of laboratory scale reactors for biomass critical chemical process studies, with most of the runs being undertaken in batch reactors [104–112] and, to a lesser extent, in continuous flow [113,114] or fluidized bed reactors [115]. The largest industrial SWG process facility is the VERENA pilot plant at Karlsruhe Institute of Technology (KIT) processing 100 kg/h. It has been designed for operation at 700°C, 35 MPa, and a maximum solids content of 20% w/w, and includes a down-flow reactor and a biomass and water preheating step with the reactor outlet stream [101,116].

In order to cut back the operation temperature and increase H_2 production, many catalysts have been developed, which can be classified into alkali-based catalysts (e.g., Na_2CO_3, K_2CO_3, or NaOH), transition metal-based catalysts (Ni and Ru), and C catalysts. All of these catalysts have been well-studied and are considered appropriate for enhancing the reaction process, increasing the H_2 yield. Despite this, alkali catalysts have been observed to cause corrosion and plugging issues, whereas transition metal- and C-based catalysts can suffer from severe deactivation [50,117].

ELECTROCHEMICAL ROUTE

Water electrolysis

Currently, the most abundant primary process for pure H_2 production is water electrolysis, and its significance is predicted to extend rapidly within the near future. Water electrolysis relies upon the movement of electrons that are supported by an external circuit. Alkaline, proton exchange membrane (PEM), and solid oxide (SO) water electrolyzers are the key electrochemical H_2 production technologies. The current density is critical in water electrolyzers; therefore, the current density in an electrolysis cell is a topic of significant research in recent years. Catalysts are utilized as one option in order to increase the current density and rate of the electrolysis reactions. Platinum (Pt) is one of the leading pure catalysts used and is applied to the surface of the electrodes within the electrolyzer. Pure catalysts can also be used throughout the electrolysis process, thanks to their high turnover rates, and are more cost-effective than the alloyed catalysts [118].

Since water electrolyzers (especially PEM water electrolyzers) are sensitive to the purity of water, desalinization, decontamination, and demineralization should be applied before the electrolysis process. If ocean water is provided to an electrolyzer, it is possible that other potential unwanted reactions such as the evolution of gases may be produced, for example, chlorine gas, Cl_2. There are many ways to prevent such reactions during the electrolysis. One straightforward method is the utilization of ion-selective membranes and reverse osmosis to clean and desalinate sea water [119,120].

ALKALINE ELECTROLYSIS CELL

In an alkaline electrolysis cell (AEC), two metallic electrodes are immersed in a liquid electrolyte. Aqueous solutions of KOH or NaOH are usually used for this purpose. The concentration of the electrolyte solution is usually up to 40% w/w to provide maximum electrical conductivity at temperatures up to 90°C [121]. The specific conductivity of a KOH solution is $54.3 \times 10^{-2}/\Omega/\text{cm}$ at 25°C [121]. Water reduction takes place at the cathode of the electrolyzer according to:

$$2H_2O + 2e^- \rightarrow H_2 + 2OH^- \left(E_C^0 = -0.828\ V\right) \tag{3.7}$$

Moreover, the oxidation of hydroxyl ions (OH^-) takes place at the anode according to:

$$2OH^- \rightarrow \frac{1}{2}O_2 + H_2O + 2e^- \left(E_A^0 = +0.401\ V\right) \tag{3.8}$$

Only water is consumed during the process. Therefore, water has to be supplied to the cell to feed the reaction and maintain the concentration of the electrolyte at an optimum. At usual operating temperatures (60–80°C), water vapor and traces of electrolyte are also carried away by the gas products. Despite this, on a dry basis, H_2 purity is usually around 99.9%.

As can be seen from Fig. 3.1, a cell separator (usually a porous material) is placed between the two electrodes to avoid the spontaneous back recombination of H_2 and O_2 into water. In a gap-cell configuration, the two electrodes are placed face-to-face, allowing a small (a few mm thick) electrolyte gap between the electrodes and the separator where gaseous H_2 and O_2 are evolved. As the current density increases, gas bubbles tend to form a continuous and highly resistive film at the surface of both electrodes. As a result, the operating current densities are limited to values up to 100 mA/cm^2. In the more-efficient zero-gap cell configuration, the two electrodes are porous and pressed against the cell

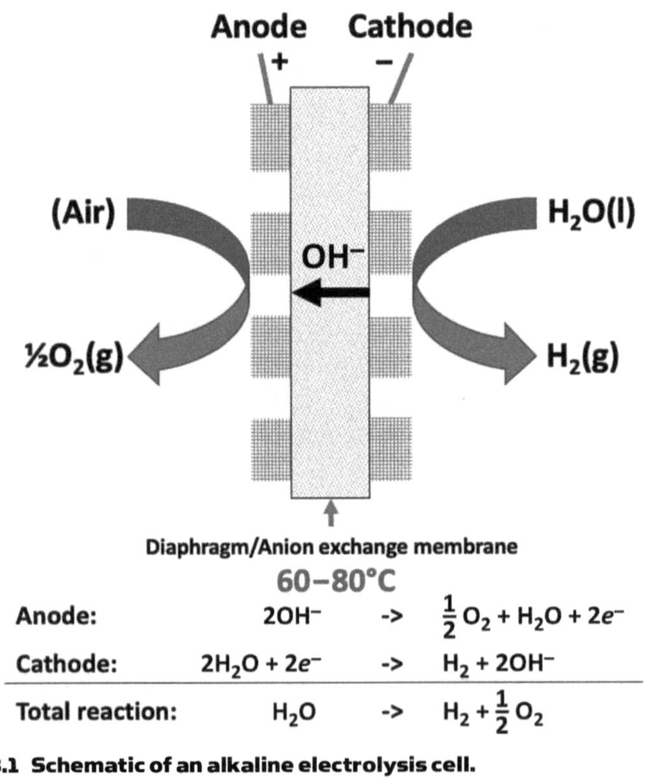

FIGURE 3.1 Schematic of an alkaline electrolysis cell.

separator. Here, the interpolar distance is lower, and gases are evolved at the back of the electrodes, allowing higher current density values to be reached.

Steel grids are usually used as electrode materials in AECs. In order to improve the charge transfer kinetics, these grids can be covered with a layer of porous nickel produced by leaching of zinc from the Ni–Zn alloy (Raney nickel). For several decades, asbestos (a natural silicate mineral) has been used for the porous diaphragm-separating anode and cathode. However, due to the toxicity of the material (the inhalation of asbestos fibers can cause lung cancer), the need for harmless materials has led to the development of a large variety of alternative solutions, in particular to the development of polymer-based composite materials [122]. It should be mentioned that the presence of a porous diaphragm generates some specific operating problems. In particular, safety issues may become critical, especially at elevated pressures, because of the possible mixing of produced gases. Moreover, gas purity tends to decrease when operating pressure is raised. Furthermore, the quality of H_2 (and O_2) in AECs is still quite low (H_2 contains impurities of O_2, water vapor, and alkali), and higher H_2 purities require additional purification steps.

Despite these drawbacks, the advantages of AECs over other water electrolysis processes are as follows:

1. They have a relatively lower initial capital requirement due to the use of cheap cell materials (electrodes and separators);
2. AEC is a proven technology with well-established operational costs;
3. The demonstration of large capacity AEC units has been achieved; and,
4. Raw water can be used directly in the process without the need for specific purification procedures.

PROTON EXCHANGE MEMBRANE ELECTROLYSIS CELL

Proton exchange membrane electrolysis cells (PEMECs) are considered the safest and most effective technology to produce H_2 from water [123]. A key component is an ion-exchange membrane. The general features of a PEMEC are shown in Fig. 3.2. Two catalyst-containing electrodes are pressed against a proton-conducting polymer electrolyte

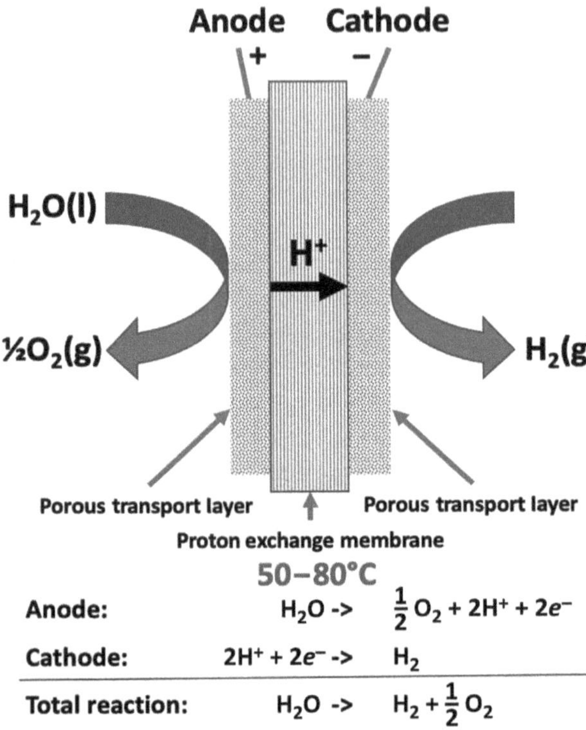

FIGURE 3.2 Schematic of a proton exchange membrane electrolysis cell.

forming a so-called membrane electrode assembly (MEA). The MEA is immersed in pure (18 MΩ/cm) water. O_2 evolution takes place at the anode:

$$H_2O \rightarrow \frac{1}{2}O_2 + 2e^- + 2H^+ \left(E^0 = +1.229\ V\ at\ 25°C\right) \tag{3.9}$$

The hydrogen ions are transported across the ion-exchange membrane and H_2 is generated at the cathode:

$$2H^+ + 2e^- \rightarrow H_2 \left(E^0 = 0\ V\ at\ 25°C\right) \tag{3.10}$$

Therefore, the overall reaction is:

$$H_2O \rightarrow H_2 + \frac{1}{2}O_2 \left(E^0 = +1.229\ V\ at\ 25°C\right) \tag{3.11}$$

Mobile species remain confined within the polymer membrane, and although they remain inside the membrane, the acidity is very high (similar to that of a 1M sulfuric acid solution). Therefore, only noble metal catalysts that can sustain such acidity are required at both the anode and cathode. PEMECs are an expensive technology; however, they were the first electrolyzers utilized for space vehicles and O_2 generation in submersible vehicles.

Today, PEMECs contain membranes made of perfluorinated phosphonic acid co-polymers (due to their relatively low resistance, high-mechanical strength, and strong chemical stability). The most widely used brand is Nafion® by DuPont de Nemours Co. (USA). These ion-exchange perfluorinated membranes are thin, elastic, and transparent films with a smooth surface. Despite their positive performance, there are some drawbacks still present with perfluorinated phosphonic acid co-polymer membranes. Moreover, during operation of a PEMEC, it has been observed that the incomplete reduction of oxygen can lead to the formation of hydrogen peroxide that can gradually degrade the membrane [124]. This causes an increase in the gas permeability during operation, therefore, decreasing the fuel cell efficiency. Additional drawbacks include the swelling and dissociation of the ion-exchange groups of the membrane when in contact with water. This results in protons freely moving within the polymer from one electrode to another.

The resistivity of such membranes is significantly larger than the resistance of aqueous solutions of alkalis (11–12 Ω/cm at 20°C and 5–6 Ω/cm at 80–90 °C). Therefore, it is necessary to use thin membranes about 100–300 μm thick to reduce ohmic losses during electrolysis. The use of thicknesses below 100 μm hurts the operation, as the permeability of gases through the membrane increases too rapidly.

There is no liquid electrolyte in PEMECs, so electrodes are held tightly against the membrane (the zero-gap configuration). In order to provide a high surface contact between the catalyst and the electrolyte, the catalyst is deposited on the surface of the membrane with the ion-exchange electrolyte [125]. Porous current collectors are pressed firmly against the catalytic layers, and adjacent electrolysis cells are separated by metallic bipolar plates.

Platinum group metals (PGM) are used as catalysts in PEMEC. Although Ru (ruthenium) has the highest catalytic activity during O_2

evolution, it is not stable in acidic media with anode potentials of +1.23 V. Iridium (Ir) is the most widely used anode catalyst currently with catalyst loadings of around 2.0 mg/cm^2. Platinum (Pt) or palladium (Pd) is the main element used as a catalyst at the cathode. Porous titanium (Ti) is generally used as the current collector. The current collectors are used to apply the potential, supply the interfaces with reagent, and to remove the gas products. The main advantages of a PEMEC are as follows:

1. There is a possibility of operating the cells at high-current densities;
2. High efficiencies can be obtained, even at high-current densities;
3. High-purity gases are produced as a result of the high-purity water used; and,
4. They have a high dynamic range.

The main drawbacks of a PEMEC are as follows:

1. Capital expenses are still very high;
2. Need for high-purity water; and,
3. Large-scale systems are not yet developed.

State-of-the-art commercial PEMECs have an enthalpy efficiency of 80% at 1.0–1.5 A/cm^2 using platinum (Pt) at the cathode and iridium oxide (IrO_2) at the anode. Commercial PEMECs can deliver high-purity H_2 (≥99.99% after dehumidification), with O_2 being the main impurity. They also have an operation lifetime in the range of 10,000+ hours, the low energy consumption of around 3.9 kWh/m^3 at 1 A/cm^2, and can deliver H_2 at high pressure directly from the cell [125]. Research has been aimed at decreasing both the operational and capital expenses of PEMEC in recent years. One such goal is the reduction of cell voltage at a constant current density. This is an important goal and may be achieved by using more efficient catalysts.

SOLID OXIDE ELECTROLYSIS CELL

The general features of solid oxide electrolysis cells (SOECs) are displayed in Fig. 3.3. Oxide-ion-conducting ceramics are used as both the solid electrolyte and cell separator. Water molecules are reduced at the cathode:

$$H_2O_{(g)} + 2e^- \rightarrow H_{2(g)} + O^{2-} \quad (3.12)$$

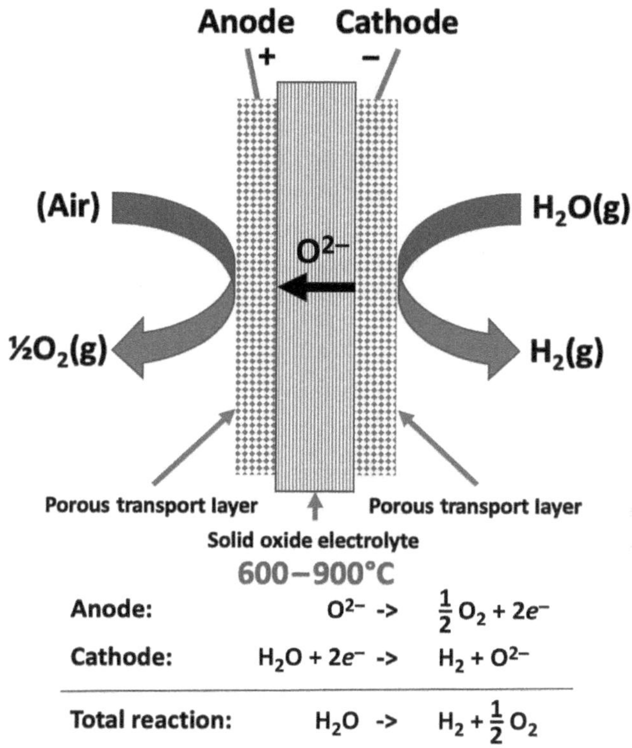

FIGURE 3.3 Schematic of a solid oxide electrolysis cell.

The resulting oxygen ions migrate to the anode, where O_2 evolves according to:

$$O^{2-} + 2e^- \rightarrow O_2 \tag{3.13}$$

SOECs usually operate in the 800–1000°C temperature range. The most typical electrolyte is zirconia stabilized with yttrium and scandium oxides [121,126]. Oxide ions are transported across the zirconia by a diffusion process. SOECs also have a very high resistivity compared to PEMECs and AECs (around 30 Ω/cm at 900–1000°C). Thin ceramic membranes that are around 30–150 μm thick are used to reduce ohmic losses. In conventional SOEC technology, the main components are stainless steel bipolar plates, and manganite-coated stabilized zirconia as the solid electrolyte.

SOEC technology development has become gradually more favorable in recent years as a tool to decarbonize the energy sector [121,127]. The possibility of operating SOECs at high-current densities and efficiencies (e.g., a current density of 3.6 A/cm^2 at +1.48 V and 950°C [128]) has driven the demand of SOECs utilization. It may also be possible to co-electrolyze water vapor and CO_2 to produce syngas (H_2 + CO) for the production of synthetic fuels. Additionally, since SOECs run at such high-operating temperatures, the electrochemical processes are entirely reversible, allowing them to operate either as a fuel or electrolysis cell.

Novel forms of electrolysis

PLASMA ARC DECOMPOSITION

Plasma is an ionizing state of matter with electrons in an excited state. Plasma has the potential to be used as a medium for high-voltage electrical transmission because of the presence of electrically charged particles. Gas (mostly CH_4) dissociates to H_2 and C as a result of the thermal plasma activity. The C is maintained in the solid fraction, whereas H_2 is readily collected as a gas. The reaction can be displayed as:

$$CH_4 \rightarrow C_{(s)} + 2H_{2(g)} \tag{3.14}$$

PHOTOELECTROLYSIS

Photoelectrolysis is the process where photocatalysts are applied to the electrodes of an electrolyzer cell. The electrolysis cell must be exposed to solar irradiation for the process to work and produce H_2, and the production of H_2 by photoelectrolysis occurs in the following steps:

1. The generation of an electron-hole;
2. A flow of electrons from the anode to the cathode;
3. The decomposition of water into H^+ and O^-;
4. The reduction of H^+ ions at the cathode to create H_2; and,
5. The separation of the produced gases, their processing, and subsequent storage.

The performance of a photoelectrolytic system depends upon the photocatalytic material (e.g., its crystalline structure, surface properties, corrosion resistance, and reactivity). There sometimes could be a trade-off between photoelectrode stability and generated energy-to-H_2-conversion potency, as the high-efficiency photoelectrodes usually have poor stability in electrolytes, whereas stable photoelectrodes exhibit poor H_2 production efficiencies.

REFERENCES

[1] Lamb JJ, Røkke G, Hohmann-Marriott MF. Chlorophyll fluorescence emission spectroscopy of oxygenic organisms at 77 K. Photosynthetica 2018;56(1):105–24.

[2] Lamb JJ, Eaton-Rye JJ, Hohmann-Marriott MF. An LED-based fluorometer for chlorophyll quantification in the laboratory and in the field. Photosynth Res 2012;114(1).

[3] Lamb J, Forfang K, Hohmann-Marriott M. A practical solution for 77 K fluorescence measurements based on LED excitation and CCD array detector. PLoS One 2015;10(7.).

[4] Kutschera A, Lamb JJ. Light meter for measuring photosynthetically active radiation. Am J Plant Sci 2018;09(12):2420–8.

[5] Lee H-S, Vermaas WFJ, Rittmann BE. Biological hydrogen production: prospects and challenges. Trends Biotechnol 2010;28(5):262–71.

[6] Allakhverdiev SI, Kreslavski VD, Thavasi V, Zharmukhamedov SK, Klimov VV, Nagata T, et al. Hydrogen photoproduction by use of photosynthetic organisms and biomimetic systems. Photochem Photobiol Sci 2009;8(2):148–56.

[7] Ghirardi ML, Posewitz MC, Maness P-C, Dubini A, Yu J, Seibert M. Hydrogenases and hydrogen photoproduction in oxygenic photosynthetic organisms. Annu Rev Plant Biol 2007;58:71–91.

[8] Ghirardi ML, King PW, Posewitz MC, Maness PC, Fedorov A, Kim K, et al. Approaches to developing biological H2-photoproducing organisms and processes. Portland Press Limited; 2005.

[9] Burgdorf T, Lenz O, Buhrke T, Van Der Linden E, Jones AK, Albracht SPJ, et al. [NiFe]-hydrogenases of Ralstonia eutropha H16: modular enzymes for oxygen-tolerant biological hydrogen oxidation. J Mol Microbiol Biotechnol 2005;10(2–4):181–96.

[10] Nishimura H, Sako Y. Purification and characterization of the oxygen-thermostable hydrogenase from the aerobic hyperthermophilic archaeon *Aeropyrum camini*. J Biosci Bioeng 2009;108(4):299–303.

[11] Vignais PM, Billoud B. Occurrence, classification, and biological function of hydrogenases: an overview. Chem Rev 2007;107(10):4206–72.

[12] Armstrong FA. Dynamic electrochemical experiments on hydrogenases. Photosynth Res 2009;102(2–3):541.

[13] Dementin S, Leroux F, Cournac L, de Lacey AL, Volbeda A, Léger C, et al. Introduction of methionines in the gas channel makes [NiFe] hydrogenase aero-tolerant. J Am Chem Soc 2009;131(29):10156–64.

[14] Cournac L, Guedeney G, Peltier G, Vignais PM. Sustained photoevolution of molecular hydrogen in a mutant of Synechocystis sp. strain PCC 6803 deficient in the type I NADPH-dehydrogenase complex. J Bacteriol 2004;186(6):1737–46.

[15] Ihara M, Nishihara H, Yoon K, Lenz O, Friedrich B, Nakamoto H, et al. Light-driven hydrogen production by a hybrid complex of a [NiFe]-hydrogenase and the cyanobacterial photosystem I. Photochem Photobiol 2006;82(3):676–82.

[16] Ihara M, Nakamoto H, Kamachi T, Okura I, Maeda M. Photoinduced hydrogen production by direct electron transfer from photosystem I cross-linked with cytochrome c 3 to [NiFe]-hydrogenase. Photochem Photobiol 2006;82(6):1677–86.

[17] Grimme RA, Lubner CE, Bryant DA, Golbeck JH. Photosystem I/molecular wire/metal nanoparticle bioconjugates for the photocatalytic production of H2. J Am Chem Soc 2008;130(20):6308–9.

[18] Beer LL, Boyd ES, Peters JW, Posewitz MC. Engineering algae for biohydrogen and biofuel production. Curr Opin Biotechnol 2009;20(3):264–71.

[19] Ghirardi ML, Mohanty P. Oxygenic hydrogen photoproduction--current status of the technology. Curr Sci 2010;98(4):499–507.

[20] Hallenbeck PC, Abo-Hashesh M, Ghosh D. Strategies for improving biological hydrogen production. Bioresour Technol 2012;110:1–9.

[21] Hallenbeck PC. Microbial paths to renewable hydrogen production. Biofuels 2011;2(3):285–302.

[22] Tsygankov AA. Hydrogen production: light-driven processes – Green Algae. In: Hallenbeck P, editors. Microbial technologies in advanced biofuels production. Boston, MA: Springer, 2012.

[23] Lee H, Krajmalinik-Brown R, Zhang H, Rittmann BE. An electron-flow model can predict complex redox reactions in mixed-culture fermentative BioH2: Microbial ecology evidence. Biotechnol Bioeng 2009;104(4):687–97.

[24] Thauer RK, Jungermann K, Decker K. Energy conservation in chemotrophic anaerobic bacteria. Bacteriol Rev 1977;41(1):100.

[25] Axley MJ, Grahame DA, Stadtman TC. Escherichia coli formate-hydrogen lyase. Purification and properties of the selenium-dependent formate dehydrogenase component. J Biol Chem 1990;265(30):18213–8.

[26] Ren N, Xing D, Rittmann BE, Zhao L, Xie T, Zhao X. Microbial community structure of ethanol type fermentation in bio-hydrogen production. Environ Microbiol 2007;9(5):1112–25.

[27] Yoshida A, Nishimura T, Kawaguchi H, Inui M, Yukawa H. Enhanced hydrogen production from glucose using ldh-and frd-inactivated Escherichiacoli strains. Appl Microbiol Biotechnol 2006;73(1):67–72.
[28] Shin J-H, Yoon JH, Ahn EK, Kim M-S, Sim SJ, Park TH. Fermentative hydrogen production by the newly isolated Enterobacter asburiae SNU-1. Int J Hydrogen Energy 2007;32(2):192–9.
[29] Li C, Fang HHP. Fermentative hydrogen production from wastewater and solid wastes by mixed cultures. Crit Rev Environ Sci Technol 2007;37(1): 1–39.
[30] Chang J-J, Chen W-E, Shih S-Y, Yu S-J, Lay J-J, Wen F-S, et al. Molecular detection of the clostridia in an anaerobic biohydrogen fermentation system by hydrogenase mRNA-targeted reverse transcription-PCR. Appl Microbiol Biotechnol 2006;70(5):598–604.
[31] Fang H, Zhang T, Liu H. Microbial diversity of a mesophilic hydrogen-producing sludge. Appl Microbiol Biotechnol 2002;58(1):112.
[32] Fang HHP, Zhang T, Li C. Characterization of Fe-hydrogenase genes diversity and hydrogen-producing population in an acidophilic sludge. J Biotechnol 2006;126(3):357–64.
[33] Levin DB, Pitt L, Love M. Biohydrogen production: prospects and limitations to practical application. Int J Hydrogen Energy 2004;29(2):173–85.
[34] Tartakovsky B, Manuel MF, Wang H, Guiot SR. High rate membrane-less microbial electrolysis cell for continuous hydrogen production. Int J Hydrogen Energy 2009;34(2):672–7.
[35] Lo Y-C, Lee K-S, Lin P-J, Chang J-S. Bioreactors configured with distributors and carriers enhance the performance of continuous dark hydrogen fermentation. Bioresour Technol 2009;100(19):4381–7.
[36] Call DF, Merrill MD, Logan BE. High surface area stainless steel brushes as cathodes in microbial electrolysis cells. Environ Sci Technol 2009;43(6): 2179–83.
[37] Call D, Logan BE. Hydrogen production in a single chamber microbial electrolysis cell lacking a membrane. Environ Sci Technol 2008;42(9):3401–6.
[38] McCarty PL, Smith DP. Anaerobic wastewater treatment. Environ Sci Technol 1986;20(12):1200–6.
[39] Wang G, Wang DIC. Elucidation of growth inhibition and acetic acid production by Clostridium thermoaceticum. Appl Environ Microbiol 1984;47(2):294–8.
[40] Williams RT, Crawford RL. Methanogenic bacteria, including an acid-tolerant strain, from peatlands. Appl Environ Microbiol 1985;50(6):1542–4.
[41] Müller V. Energy conservation in acetogenic bacteria. Appl Environ Microbiol 2003;69(11):6345–53.
[42] Müller B, Sun L, Schnürer A. First insights into the syntrophic acetate-oxidizing bacteria–a genetic study. Microbiol Open 2013;2(1):35–53.
[43] de Bok FAM, Stams AJM, Dijkema C, Boone DR. Pathway of Propionate Oxidation by a Syntrophic Culture of Smithella propionica andMethanospirillum hungatei. Appl Environ Microbiol 2001;67(4):1800–4.

[44] Penner SS. Steps toward the hydrogen economy. Energy 2006;31(1):33–43.
[45] Ni M, Leung DYC, Leung MKH. A review on reforming bio-ethanol for hydrogen production. Int J Hydrogen Energy 2007;32(15):3238–47.
[46] Wang D, Czernik S, Chornet E. Production of hydrogen from biomass by catalytic steam reforming of fast pyrolysis oils. Energy Fuels 1998;12(1): 19–24.
[47] Tanksale A, Beltramini JN, Lu GM. A review of catalytic hydrogen production processes from biomass. Renew Sustain Energy Rev 2010;14(1):166–82.
[48] Steinfeld A. Solar thermochemical production of hydrogen—a review. Sol energy 2005;78(5):603–15.
[49] Ni M, Leung MKH, Leung DYC, Sumathy K. A review and recent developments in photocatalytic water-splitting using TiO2 for hydrogen production. Renew Sustain Energy Rev 2007;11(3):401–25.
[50] Guo Y, Wang SZ, Xu DH, Gong YM, Ma HH, Tang XY. Review of catalytic supercritical water gasification for hydrogen production from biomass. Renew Sustain Energy Rev 2010;14(1):334–43.
[51] Choudhary TV, Sivadinarayana C, Goodman DW. Catalytic ammonia decomposition: COx-free hydrogen production for fuel cell applications. Catal Letters 2001;72(3–4):197–201.
[52] Nikolaidis P, Poullikkas A. A comparative overview of hydrogen production processes. Renew Sustain energy Rev 2017;67:597–611.
[53] Rostrup-Nielsen JR. Catalytic steam reforming. In: Anderson JR, Boudart M, editors. Catalysis. Catalysis (Science and Technology), vol. 5. Berlin, Heidelberg: Springer, 1984.
[54] Mazanec TJ, Cable TL. Inventors; Standard Oil Co, assignee. Process for the partial oxydation of hydrocarbons. United States patent US 5,714,091, 1998.
[55] Chen HL, Lee HM, Chen SH, Chao Y, Chang MB. Review of plasma catalysis on hydrocarbon reforming for hydrogen production—interaction, integration, and prospects. Appl Catal B Environ 2008;85(1–2):1–9.
[56] Dicks AL. Hydrogen generation from natural gas for the fuel cell systems of tomorrow. J Power Sources 1996;61(1–2):113–24.
[57] Jenkins HDB. Thermodynamics at a Glance. Oxford, UK: Blackwell Publishing; 2007.
[58] Rostrup-Nielsen JR, Rostrup-Nielsen T. Large-scale hydrogen production. Cattech 2002;6(4):150–9.
[59] Jechura J. Hydrogen from natural gas via steam methane reforming (SMR), vol. 4. CO, USA: Color Sch Mines Golden, 2015. Available from: http://inside.mines.edu/~jjechura/EnergyTech/07_Hydrogen_from_SMR.pdf
[60] Haryanto A, Fernando S, Murali N, Adhikari S. Current status of hydrogen production techniques by steam reforming of ethanol: a review. Energy Fuels 2005;19(5):2098–106.
[61] Chen G, Tao J, Liu C, Yan B, Li W, Li X. Hydrogen production via acetic acid steam reforming: a critical review on catalysts. Renew Sustain Energy Rev 2017;79: 1091–8.

[62] Liguras DK, Kondarides DI, Verykios XE. Production of hydrogen for fuel cells by steam reforming of ethanol over supported noble metal catalysts. Appl Catal B Environ 2003;43(4):345–54.
[63] Sehested J, Jacobsen CJH, Rokni S, Rostrup-Nielsen JR. Activity and stability of molybdenum carbide as a catalyst for CO_2 reforming. J Catal 2001;201(2): 206–12.
[64] Hawkins G. Steam Reforming Practical Operation. Chicago: GBH Enterp Ltd; 2013.
[65] Uemiya S. Brief review of steam reforming using a metal membrane reactor. Top Catal 2004;29(1–2):79–84.
[66] De Falco M, Marrelli L, Iaquaniello G. Membrane reactors for hydrogen production processes. Vol. 264. Springer; 2011.
[67] Adris AM, Elnashaie S, Hughes R. A fluidized bed membrane reactor for the steam reforming of methane. Can J Chem Eng 1991;69(5):1061–70.
[68] Chen Z, Yan Y, Elnashaie SSEH. Novel circulating fast fluidized-bed membrane reformer for efficient production of hydrogen from steam reforming of methane. Chem Eng Sci 2003;58(19):4335–49.
[69] Steinberg M, Cheng HC. Modern and prospective technologies for hydrogen production from fossil fuels. Int J Hydrogen Energy 1989;14(11): 797–820.
[70] Clarke SH, Dicks AL, Pointon K, Smith TA, Swann A. Catalytic aspects of the steam reforming of hydrocarbons in internal reforming fuel cells. Catal Today 1997;38(4): 411–23.
[71] Morosuk T, Tsatsaronis G, Koroneos C. Environmental impact reduction using exergy-based methods. J Clean Prod 2016;118:118–23.
[72] Angeli SD, Monteleone G, Giaconia A, Lemonidou AA. State-of-the-art catalysts for CH_4 steam reforming at low temperature. Int J Hydrogen Energy 2014;39(5): 1979–97.
[73] Dahl PJ, Christensen TS, Winter-Madsen S, King SM. Proven autothermal reforming technology for modern large-scale methanol plants. In: Nitrogen+ Syngas. Int Proc Conf Exhib 2014;2–3.
[74] Park Y, Namioka T, Sakamoto S, Min T, Roh S, Yoshikawa K. Optimum operating conditions for a two-stage gasification process fueled by polypropylene by means of continuous reactor over ruthenium catalyst. Fuel Process Technol 2010;91(8) :951–7.
[75] Wu C, Williams PT. Pyrolysis–gasification of plastics, mixed plastics and real-world plastic waste with and without Ni–Mg–Al catalyst. Fuel 2010;89(10): 3022–32.
[76] Barbarias I, Lopez G, Alvarez J, Artetxe M, Arregi A, Bilbao J, et al. A sequential process for hydrogen production based on continuous HDPE fast pyrolysis and in-line steam reforming. Chem Eng J 2016;296:191–8.
[77] Barbarias I, Lopez G, Artetxe M, Arregi A, Santamaria L, Bilbao J, et al. Pyrolysis and in-line catalytic steam reforming of polystyrene through a two-step reaction system. J Anal Appl Pyrolysis 2016;122:502–10.

[78] Arregi A, Lopez G, Amutio M, Barbarias I, Bilbao J, Olazar M. Hydrogen production from biomass by continuous fast pyrolysis and in-line steam reforming. RSC Adv 2016;6(31):25975–85.

[79] Lan P, Xu Q, Zhou M, Lan L, Zhang S, Yan Y. Catalytic Steam Reforming of Fast Pyrolysis Bio-Oil in Fixed Bed and Fluidized Bed Reactors. Chem Eng Technol 2010;33(12):2021–8.

[80] Erkiaga A, Lopez G, Barbarias I, Artetxe M, Amutio M, Bilbao J, et al. HDPE pyrolysis-steam reforming in a tandem spouted bed-fixed bed reactor for H2 production. J Anal Appl Pyrolysis 2015;116:34–41.

[81] Zhang W. Automotive fuels from biomass via gasification. Fuel Process Technol 2010;91(8):866–76.

[82] Bulushev DA, Ross JRH. Catalysis for conversion of biomass to fuels via pyrolysis and gasification: a review. Catal today 2011;171(1):1–13.

[83] Heidenreich S, Foscolo PU. New concepts in biomass gasification. Prog energy Combust Sci 2015;46:72–95.

[84] Sikarwar VS, Zhao M, Clough P, Yao J, Zhong X, Memon MZ, et al. An overview of advances in biomass gasification. Energy Environ Sci 2016;9(10): 2939–77.

[85] Miao Z, Shastri Y, Grift TE, Hansen AC, Ting KC. Lignocellulosic biomass feedstock transportation alternatives, logistics, equipment configurations, and modeling. Biofuels Bioprod Biorefining 2012;6(3):351–62.

[86] Parthasarathy P, Narayanan KS. Hydrogen production from steam gasification of biomass: influence of process parameters on hydrogen yield–a review. Renew energy 2014;66:570–9.

[87] Mohan D, Pittman CU, Steele PH. Pyrolysis of wood/biomass for bio-oil: a critical review. Energy Fuels 2006;20(3):848–89.

[88] Di Blasi C. Modeling chemical and physical processes of wood and biomass pyrolysis. Prog energy Combust Sci 2008;34(1):47–90.

[89] Kan T, Strezov V, Evans TJ. Lignocellulosic biomass pyrolysis: A review of product properties and effects of pyrolysis parameters. Renew Sustain Energy Rev 2016; 57:1126–40.

[90] Devi L, Ptasinski KJ, Janssen FJJG, van Paasen SVB, Bergman PCA, Kiel JHA. Catalytic decomposition of biomass tars: use of dolomite and untreated olivine. Renew energy 2005;30(4):565–87.

[91] Palma CF. Modelling of tar formation and evolution for biomass gasification: a review. Appl Energy 2013;111:129–41.

[92] Erkiaga A, Lopez G, Amutio M, Bilbao J, Olazar M. Steam gasification of biomass in a conical spouted bed reactor with olivine and γ-alumina as primary catalysts. Fuel Process Technol 2013;116:292–9.

[93] Peng WX, Wang LS, Mirzaee M, Ahmadi H, Esfahani MJ, Fremaux S. Hydrogen and syngas production by catalytic biomass gasification. Energy Convers Manag 2017;135:270–3.

[94] Sansaniwal SK, Rosen MA, Tyagi SK. Global challenges in the sustainable development of biomass gasification: an overview. Renew Sustain Energy Rev 2017;80:23–43.

[95] Ollero P, Serrera A, Arjona R, Alcantarilla S. The CO_2 gasification kinetics of olive residue. Biomass Bioener 2003;24(2):151–61.

[96] Lopez G, Alvarez J, Amutio M, Arregi A, Bilbao J, Olazar M. Assessment of steam gasification kinetics of the char from lignocellulosic biomass in a conical spouted bed reactor. Energy 2016;107:493–501.

[97] Ahmed II, Gupta AK. Kinetics of woodchips char gasification with steam and carbon dioxide. Appl Energy 2011;88(5):1613–9.

[98] Pereira EG, da Silva JN, de Oliveira JL, Machado CS. Sustainable energy: a review of gasification technologies. Renew Sustain Energy Rev 2012;16(7): 4753–62.

[99] Nanda S, Isen J, Dalai AK, Kozinski JA. Gasification of fruit wastes and agro-food residues in supercritical water. Energy Convers Manag 2016;110: 296–306.

[100] Norouzi O, Safari F, Jafarian S, Tavasoli A, Karimi A. Hydrothermal gasification performance of Enteromorpha intestinalis as an algal biomass for hydrogen-rich gas production using Ru promoted Fe–Ni/γ-Al2O3 nanocatalysts. Energy Convers Manag 2017;141:63–71.

[101] Yakaboylu O, Harinck J, Smit K, De Jong W. Supercritical water gasification of biomass: a literature and technology overview. Energies 2015;8(2): 859–94.

[102] Matsumura Y, Minowa T, Potic B, Kersten SRA, Prins W, van Swaaij WPM, et al. Biomass gasification in near-and super-critical water: status and prospects. Biomass Bioener 2005;29(4):269–92.

[103] Calzavara Y, Joussot-Dubien C, Boissonnet G, Sarrade S. Evaluation of biomass gasification in supercritical water process for hydrogen production. Energy Convers Manag 2005;46(4):615–31.

[104] Safari F, Salimi M, Tavasoli A, Ataei A. Non-catalytic conversion of wheat straw, walnut shell and almond shell into hydrogen rich gas in supercritical water media. Chinese J Chem Eng 2016;24(8):1097–103.

[105] Yanik J, Ebale S, Kruse A, Saglam M, Yüksel M. Biomass gasification in supercritical water: part 1. Effect of the nature of biomass. Fuel 2007;86(15): 2410–5.

[106] Lu YJ, Guo LJ, Ji CM, Zhang XM, Hao XH, Yan QH. Hydrogen production by biomass gasification in supercritical water: a parametric study. Int J Hydrogen Energy 2006;31(7):822–31.

[107] Lu Y, Guo L, Zhang X, Ji C. Hydrogen production by supercritical water gasification of biomass: explore the way to maximum hydrogen yield and high carbon gasification efficiency. Int J Hydrogen Energy 2012;37(4):3177–85.

[108] Yanik J, Ebale S, Kruse A, Saglam M, Yüksel M. Biomass gasification in supercritical water: II. Effect of catalyst. Int J Hydrogen Energy 2008;33(17): 4520–6.

[109] Ding N, Azargohar R, Dalai AK, Kozinski JA. Catalytic gasification of cellulose and pinewood to H_2 in supercritical water. Fuel 2014;118: 416–25.

[110] Jin H, Lu Y, Guo L, Zhang X, Pei A. Hydrogen production by supercritical water gasification of biomass with homogeneous and heterogeneous catalyst. Adv Condens Matter Phys 2014;69:9. doi:10.1155/2014/160565.
[111] Elif D, Nezihe A. Hydrogen production by supercritical water gasification of fruit pulp in the presence of Ru/C. Int J Hydrogen Energy 2016;41(19): 8073–83.
[112] Osada M, Yamaguchi A, Hiyoshi N, Sato O, Shirai M. Gasification of sugarcane bagasse over supported ruthenium catalysts in supercritical water. Energy Fuels 2012;26(6):3179–86.
[113] Madenoğlu TG, Boukis N, Sağlam M, Yüksel M. Supercritical water gasification of real biomass feedstocks in continuous flow system. Int J Hydrogen Energy 2011;36(22):14408–15.
[114] Antal MJ Jr, Allen SG, Schulman D, Xu X, Divilio RJ. Biomass gasification in supercritical water. Ind Eng Chem Res 2000;39(11):4040–53.
[115] Lu YJ, Jin H, Guo LJ, Zhang XM, Cao CQ, Guo X. Hydrogen production by biomass gasification in supercritical water with a fluidized bed reactor. Int J Hydrogen Energy 2008;33(21):6066–75.
[116] Kruse A. Supercritical water gasification. Biofuels, Bioprod Biorefining Innov a Sustain Econ 2008;2(5):415–37.
[117] Azadi P, Farnood R. Review of heterogeneous catalysts for sub-and supercritical water gasification of biomass and wastes. Int J Hydrogen Energy 2011;36(16): 9529–41.
[118] Karunadasa HI, Chang CJ, Long JR. A molecular molybdenum-oxo catalyst for generating hydrogen from water. Nature 2010;464(7293):1329.
[119] El-Bassuoni A-M, Sheffield JW, Veziroglu TN. Hydrogen and fresh water production from sea water. Int J Hydrogen Energy 1982;7(12):919–23.
[120] Ni M, Leung MKH, Sumathy K, Leung DYC. Potential of renewable hydrogen production for energy supply in Hong Kong. Int J Hydrogen Energy 2006;31(10): 1401–12.
[121] Millet P, Grigoriev S. Water electrolysis technologies. Renew Hydrog Technol Prod Purification, Storage, Appl Saf 2013;19–41.
[122] Rosa VM, Santos MBF, Da Silva EP. New materials for water electrolysis diaphragms. Int J Hydrogen Energy 1995;20(9):697–700.
[123] Grigoriev SA, Porembsky VI, Fateev VN. Pure hydrogen production by PEM electrolysis for hydrogen energy. Int J Hydrogen Energy 2006;31(2):171–5.
[124] Kinumoto T, Inaba M, Nakayama Y, Ogata K, Umebayashi R, Tasaka A, et al. Durability of perfluorinated ionomer membrane against hydrogen peroxide. J Power Sources 2006;158(2):1222–8.
[125] Grigoriev SA, Millet P, Fateev VN. Evaluation of carbon-supported Pt and Pd nanoparticles for the hydrogen evolution reaction in PEM water electrolysers. J Power Sources 2008;177(2):281–5.
[126] O'brien JE, Stoots CM, Herring JS, Hartvigsen J. Hydrogen production performance of a 10-cell planar solid-oxide electrolysis stack. J Fuel Cell Sci Technol 2006;3(2): 213–9.

[127] Stoots C, O'Brien JE, Hawkes GL, Herring JS, Hartvigsen JJ. High temperature steam and carbon dioxide electrolysis experiments at INL. In: Workshop on High Temperature Electrolysis, Roskilde, Denmark. 2006.
[128] Jensen SH, Larsen PH, Mogensen M. Hydrogen and synthetic fuel production from renewable energy sources. Int J Hydrogen Energy 2007;32(15): 3253–7.

Chapter | four

Emerging Technology for Hydrogen and Bioenergy Production

Jacob J. Lamb[a], Erling Rytter[b], Magne Hillestad[b], Kristian M. Lien[c], Odne S. Burheim[d], Bruno G. Pollet[d]

[a]*Department of Electronic Systems, Department of Energy and Process Engineering, ENERSENSE, Faculty of Information Technology and Electrical Engineering, NTNU, Trondheim, Norway;* [b]*Department of Chemical Engineering, ENERSENSE, Faculty of Natural Sciences, NTNU, Trondheim, Norway;* [c]*Department of Energy and Process Engineering, ENERSENSE, Faculty of Engineering, NTNU, Trondheim, Norway;* [d]*Department of Energy and Process Engineering, ENERSENSE & NTNU Team Hydrogen, Faculty of Engineering, NTNU, Trondheim, Norway*

CHAPTER OUTLINE

ABSTRACT

Nuclear heat, industrial heat, waste heat recovered from power plants, concentrated solar heat, heat resulting from municipal waste incineration, and geothermal heat can be listed as sustainable thermal energy sources. Hybrid thermochemical cycles operate at lower temperatures compared to thermally driven water-splitting cycles. As a result, nuclear heat, industrial heat, waste heat recovered from power

Hydrogen, Biomass and Bioenergy. http://dx.doi.org/10.1016/B978-0-08-102629-8.00004-9

plants, concentrated solar heat, heat resulting from municipal waste incineration, and geothermal heat can be listed as sustainable thermal energy sources for thermochemical cycles. Since hybrid cycles operate at lower temperatures, other sustainable thermal sources apart from solar, high-temperature nuclear, and biomass combustion (such as recovered waste heat from nuclear and geothermal facilities) can be used to drive thermochemical processes. This chapter gives an overview of such emerging technologies for hydrogen and bioenergy production.

HYBRID THERMOCHEMICAL HYDROGEN PRODUCTION ROUTES

Hybrid thermochemical cycles operate at lower temperatures compared to thermally driven water-splitting cycles mentioned in thermochemical water splitting (Chapter 3).

External energy needs of the individual electrochemical reactions are met by thermal and electrical energies. Since hybrid cycles operate at lower temperatures, other sustainable thermal sources apart from solar, high-temperature nuclear, and biomass combustion (such as recovered waste heat from nuclear and geothermal facilities) can be used to drive the involved processes.

The Cu–Cl cycle is proposed as an outstanding hybrid cycle as investigated in the literature. The operating temperature of this cycle does not exceed 550°C. Among different types of Cu–Cl cycles, the "five-step" version is the most studied one. This version is composed of three thermally driven chemical reactions, one electrochemical reaction, and one physical drying step. In Cu–Cl cycles, the thermal energy source is used partially to drive the cycle directly and partially to generate the required electricity. The significant advantage of Cu–Cl cycles is H_2 generation from low-grade temperature sources, especially those that can be considered as sustainable thermal energy. Nuclear heat, industrial heat, waste heat recovered from power plants, concentrated solar heat, heat resulting from municipal waste incineration, and geothermal heat can be listed as sustainable thermal energy sources.

Renewable energy generation

It is clean and energy saving to generate power via geothermal energy. Pilot and commercial geothermal power plants have been built in many countries (e.g., Iceland and New Zealand). However, there still exists issues such as the distribution of geothermal resources. Generally, wind energy is converted to electrical energy using a wind turbine. Power generation from wind systems mainly consists of a wind turbine, an inverter, a controller, and other electrical devices. Solar radiation is abundant, dominant, and in many respects, the primary long-term sustainable energy option for all human activity. Generally, renewable energy generation systems can either be off-grid or grid-connected depending upon the local load and the grid distribution. For an off-grid system, an energy storage system is needed to handle power fluctuations due to the intermittent and location-specific nature of geothermal, solar, and wind energy.

Renewable energy can be transformed to more convenient energy carriers by biochemical, photochemical, or thermal means. Water (H_2O) can be split to H_2 and O_2 with optional subsequent uploading of all types of carbon-containing feedstock (including biomass), utilizing high-temperature-concentrated solar radiation. The term uploading refers to the process whereby the original carbonaceous material is converted into intermediates or products with enhanced energy content compared to the feed material. The principle is shown in Fig. 4.1.

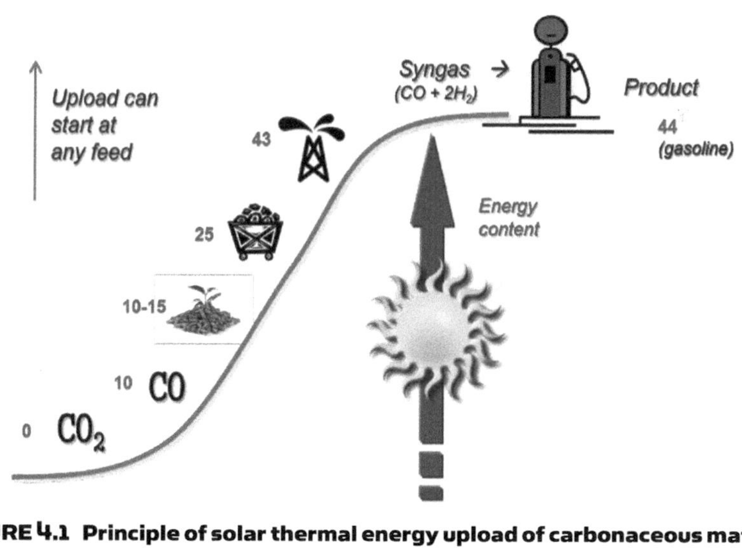

FIGURE 4.1 Principle of solar thermal energy upload of carbonaceous material with H_2. *Red* numbers: Energy of combustion (LHV; kJ/g).

A particularly suitable intermediate is synthesis gas (syngas) (i.e., a mixture of H_2 and CO), which can be converted to other energy carriers by conventional means. The final products encompass H_2, gasoline, diesel, methanol, ammonia, and CH_4 gas as well as a petrochemical feedstock for further conversion to a variety of products, including plastics.

Concentrated solar power (CSP) plants are based upon a vast number of mirrors collected in heliostats. A central receiver (i.e., a reactor for water splitting in the present case) is surrounded by an array of automated mirrors. These mirrors are designed to track The Sun's motion throughout the day and concentrate the solar radiation on a central tower containing collector mirrors that direct the radiation to the aperture of the chemical reactor [1].The principle is depicted in Fig. 4.2, including notations for heat fluxes. CSP plants can convert an abundance of solar energy while The Sun is shining. Storage of energy for dark periods can be provided by a thermal heat sink (e.g., water or molten salts).

The efficiency (defined as $\eta_f = Q_c/Q_s$) from the incoming solar radiation to the heliostat and further by the collector to the reactor is approximately 0.6 for a typical heliostat and reactor design. Up to 70% of this energy can be absorbed by the water-splitting reaction ($\eta_{abs} = Q_\varepsilon/Q_c$; reduction of oxide), giving an overall efficiency of 40%.

Novel hybrid renewable energy systems that integrated solar, wind, and geothermal energy are also possible. Three types of subsystems, namely, a solar photovoltaic (PV) system, a wind turbine system, and a geothermal system, are independently connected to an energy storage

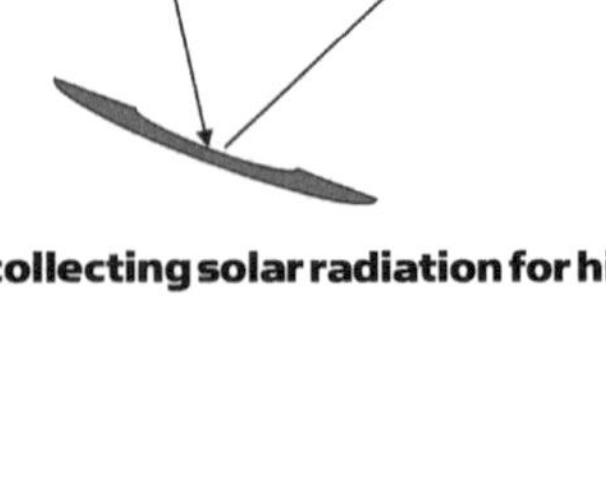

FIGURE 4.2 Principle of collecting solar radiation for high-temperature reactions.

system used for storing surplus electricity and balancing the load. The hybrid multi-renewable energy system is potentially more reliable and cost-effective.

Nuclear energy generation

Nuclear power plants are capable of providing consistent and low-carbon electricity (excluding embedded carbon emissions). These plants do not discharge any greenhouse gas during operation and comparatively require a much smaller land footprint [2]. Presently, nuclear power plants fulfill over 11% of the world's electricity demand, and there are more than 60 nuclear reactors under construction in various countries at present [3,4]. Despite being a proven technology and an attractive option for power generation in order to progress toward a green world, nuclear reactors have not progressed rapidly and are not accepted widely because of various issues and challenges. Anti-nuke policies and public opinion over safety and waste disposal are the largest hurdles for its expansion [2]. Despite its controversial reputation, nuclear power has been described as sustainable, efficient, and reliable, and various studies argued that it must be part of the energy mix [5,6].

Depending upon various factors (e.g., geographical, economical, and desired form of output), different renewable energy sources may be coupled to form a nuclear-renewable hybrid energy system. An integrated nuclear-renewable energy system could be operated by electricity demand; that is, if the sum of their electrical output is enough to satisfy the variation of electricity need. Moreover, H_2 production units could operate on surplus power generated during the peak periods from renewable sources, while nuclear power serves as the baseline energy source.

Thermochemical hydrogen production

THERMOCHEMICAL CYCLES

All of the above can be achieved by uploading CO_2, as a final solution, or in a transient scenario by starting with biomass. The technology takes concentrated radiation (heat) and abstracts an oxygen atom from a gaseous water molecule. Before the hydrolysis of steam, an oxide is activated by expelling oxygen atoms from the crystal lattice at the given

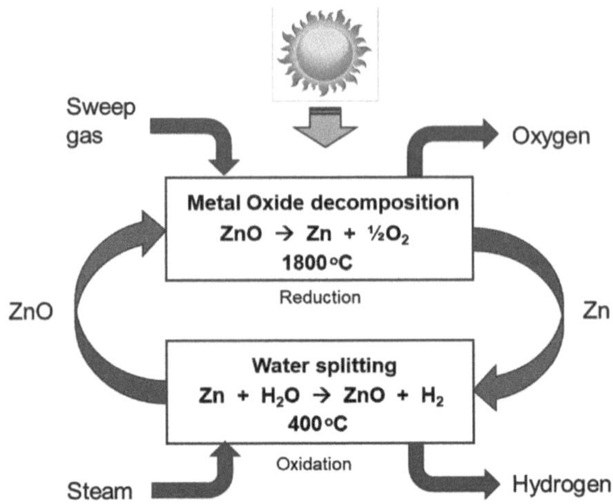

FIGURE 4.3 Solar thermal zinc/zinc oxide cycle for water splitting.

serverity conditions. The principle is exemplified by the Zn-ZnO cycle for water splitting shown in Fig. 4.3. The figure illustrates cycling of the solid material as Zn or ZnO, respectively, between a reduction and an oxidation reactor [7]. In this case, the solids are transported as spherical particles, and the reactors are of the fluidized- or bubbling-bed types. A viable alternative is to cycle the gases instead; that is, keeping one reactor but feed intermittently with steam and inert gas that sweeps out the produced O_2. We notice that the oxide decomposition temperature is exceedingly high (around 1800°C). Further, there is thermal cycling between the two steps of 1400°C.

The aspect of solar thermal upload by water splitting to H_2 is often referred to as a Solar Thermochemical Cycle (STC) for H_2 production. Many different cycles have been studied, and a particularly popular one is based on unstoichiometric ceria (CeO_x). Within the EU framework programs, there has been a series of projects under the acronym HYDROSOL [8]. In particular, the HYDROSOL-3D project focused on reactor design and construction for high-temperature solar H_2 production using the Ce^{IV}-oxide → Ce^{III}-oxide system implemented in a monolith-type reactor. A 750-kW system was operated in the 2014–16 project.

Excellent reviews and evaluations of the investigated systems have been reported [9,10]. In the SANDIA report, initial screening reduced 350 possible cycles to 9 cycles in 5 reaction classes that were analyzed in more detail. These cycles were of widely different maturity with investigations ranging from 30 years to less than a year. All systems exhibit technical challenges of some sort (e.g., exceedingly high temperature, vast temperature difference between reduction and oxidation, need for regeneration, attrition and stability of solids, and low efficiency). The least mature cycle is the ferrite/hercynite cycle, which is described further. However, it was concluded that this cycle is particularly promising in terms of less severe process conditions and high-yield potential. A few groups (ETH and University of Colorado Boulder) have been actively studying the hercynite cycle and some references are given in a review on oxygen exchange materials for STC [11].

THE FERRITE AND HERCYNITE CYCLES

In the cobalt ferrite-based cycle, the reduction reaction is:

$$CoFe_2O_4 \rightarrow 2CoO + FeO + \frac{1}{2}O_{2(g)} \qquad (4.1)$$

Therefore, reducing iron from valence state III in the ferrite to valence state II in divalent iron oxide. In the hercynite cycle, ferrite[a] is deposited on support of alumina. The alumina system exhibits the unique property of lowering the decomposition temperature of cobalt ferrite ($CoFe_2O_4$) by up to 300 degrees, yielding hercynite ($FeAl_2O_4$), by the following reaction:

$$Co^{II}Fe_2^{III}O_4 + 3Al_2O_2 \rightarrow 2Fe^{II}Al_2O_4 + Co^{II}Al_2O_4 + \frac{1}{2}O_{2(g)} \qquad (4.2)$$

Reduction of Fe^{III} to Fe^{II} is followed by expelling oxygen from the solid crystal lattice, and the co-current reaction with alumina to give iron- and cobalt-aluminate, making the system stoichiometric. Upon

[a] Ferrite refers to minerals where transition metal ions are substituted for divalent iron in magnetite (Fe_3O_4) giving $M_xFe_{3-x}O_4$ spinel.

reoxidation with H_2O, ferrite and alumina are reformed by the reaction:

$$2Fe^{II}Al_2O_4 + Co^{II}Al_2O_4 + H_2O_{(g)} \rightarrow Co^{II}Fe_2^{III}O_4 + 3Al_2O_3 + H_{2(g)} \quad (4.3)$$

(In equation 4.3, $H_2O_{(g)}$ can be replaced with ½ $CO_{2(g)}$, resulting in $CO_{(g)}$ production instead of $H_{2(g)}$.)

Although it has been claimed that one of the advantages of the ferrite cycles is that the reactions are stoichiometric, this has been disputed in recent work [9]. In any case, there certainly are solid-state microkinetics and thermodynamics needed to elucidate mechanisms and the potential of this class of materials. An H_2 yield of 600 $\mu mol/g_{ferrite}$ (4 times greater than for the pure cobalt-ferrite system), was found at a temperature as moderate as 1200°C [12]. The applied conditions are several hundred degrees lower than for most other oxide systems studied.

Interestingly, Muhich et al. showed that the hercynite cycle could be operated isothermally at 1350°C with high efficiency [13]. They claim six times higher productivity on a total weight basis (12 times on an active material basis) than the ceria cycles often regarded as the present state-of-the-art. The reduction and oxidation reactions are driven by changes in the oxygen partial pressure rather than conventional temperature cycling. This suggests from these conditions that irreversible heat losses can be significantly reduced, and thermal shock concerns can be practically eliminated. However, a theoretical study indicates that some temperature cycling is beneficial [14].

SOLAR THERMAL PROCESS

Several processes for making H_2 or syngas by STC have been analyzed. All of these face challenges with materials stability and corrosion at high temperatures. Special attention has to be given to the design of the solar reactor itself. Also, advances in ceramic heat exchangers are required to cool produced H_2, O_2, and CO (in cases where CO_2 is split to make CO for further fuels synthesis [15]). Effective heat integration is significantly essential when operating at these extreme temperatures. This involves heating of feed streams and utilizing the energy in hot streams.

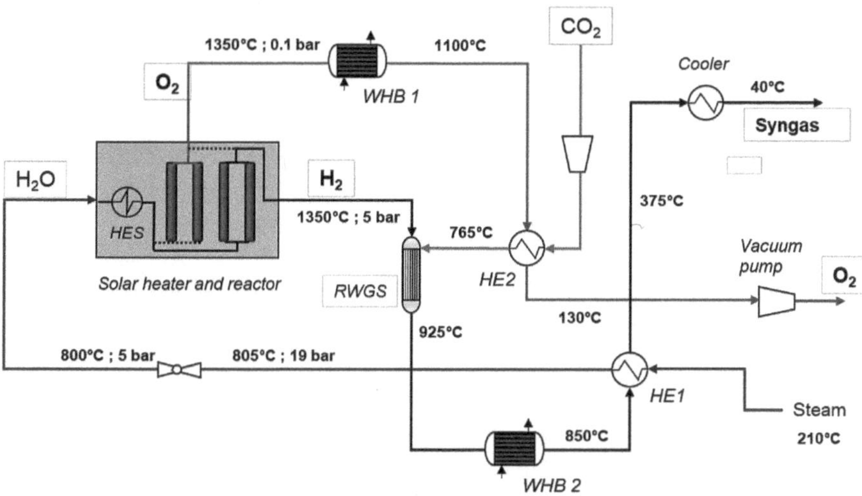

FIGURE 4.4 Solar thermal process for generation of H_2 and synthesis gas.

A developed STC process concept for producing syngas is shown in Fig. 4.4 [16]. Syngas can be used to produce fuels or methanol, as described in Chapter 6. The unique feature of this process design is that there is no need to split CO_2 by an STC, but CO_2 is instead used as a coolant of the produced H_2 by chemical quenching. Chemical quenching means that a reactant is introduced to favor an endothermic reaction with an associated reduction of temperature. Also, the lowered temperature of the introduced reactant strengthens the effect. Of course, it is possible to produce H_2 directly without further reaction with CO_2 to syngas, but the high temperatures generate several technical and material stability challenges that are relaxed by the combined process. Despite this, let us follow the produced H_2.

H_2 at 1350°C is quenched to 925°C by CO_2 that has been preheated to 765°C. Quenching takes place in a reverse water-gas-shift reactor (RWGS; see details in Chapter 6). A temperature of 900–1000°C is sufficiently low to allow further heat exchange with boiling water in a conventional waste-heat-boiler (WHB), which makes steam for process purposes or electricity generation. Hot syngas is then used to preheat steam in the first heat exchanger (HE1) to 805°C before the solar reactor. Further heating of steam before water splitting takes place with solar radiation in the solar heat exchanger (HES) as the front end of the reactor.

The produced O_2 deserves a comment. First, notice that there is no flushing of the O_2; however, the O_2 is removed by using a vacuum pump. A vacuum secures a sufficient driving force for the reduction of the oxide $CoFe_2O_4$. Benefits of not having a sweep gas are several. The O_2 can be used for processes where dilution with N_2 (or another inert gas) can be avoided.

Furthermore, costly production of inert gas (e.g., an air-separation unit) is not required. O_2 at these reaction temperatures is highly reactive, and design with ceramic tubes to a partly ceramic WHB is required. At 1100°C, steel can be employed without softening for use in the second heat exchanger (HE2), used to preheat the CO_2. Having oxidative gases in this heat exchanger prevents corrosion, specifically metal dusting.

This example shows that the operating processes at solar thermal temperatures are not trivial, but doable. Further developments in STC materials, reactor design, heat-exchanger design, and process optimization are required. We have not addressed in detail what is happening when one shifts from reductive to oxidative conditions in the STC reactor. Flushing out the gases from the previous step is needed, partial pressures and solid compositions change with reaction time, and there is a gradient along the reactor length that needs to be considered.

Electrochemical hydrogen production

There are many different types of electrochemical water electrolyzers for producing H_2. Of these, the three most common are alkaline, proton exchange membrane, and solid oxide electrolysis cells (as described in Chapter 3).

ALKALINE ELECTROLYSIS CELL

Alkaline electrolysis cells (AECs) are considered one of the primary processes for the electrochemical water-splitting reaction. Military applications related to the use of H_2 isotopes boosted their development in the 20th century. The first plants for the electrolysis of heavy water and the production of deuterium were built in Norway. Nowadays, several companies are manufacturing AECs for the production of high-grade H_2 (e.g., NEL Hydrogen; a former department of Norsk Hydro Co., Norway). NEL Hydrogen (Norway) is a world-leading producer of electrolytic systems. Their electrolyzers can deliver 50–485 Nm^3 H_2 per hour (when operating at atmospheric pressure), with a purity of 99.9%. Energy consumption currently is usually 4.1–4.3 kWh per Nm^3 H_2 (with current densities up to 0.3 A/cm^2). AECs can be manufactured for a variety of applications, and the global H_2 demand is growing fast. The major industrial H_2 gas-consuming industries are currently:

1. Electrical power generator cooling;
2. Manufacturing of semiconductors;
3. The production of flat screens for TVs and computers;
4. Heat-treatment plants; and,
5. Scientific laboratories.

Other industrial users of H_2 are food processing, glass manufacturing, welding, heat-treating, and meteorology industries. In recent decades, the use of H_2 as an energy carrier is gaining more traction. This will open up new applications allowing the management of smart grids for more energy flexibility, renewable energy storage, and transport applications.

AECs are considered as a mature technology. There are several MW industrial AECs that are already implemented and used in the industry sector, leading to the development of AECs that can produce approximately 60 kg H_2 per hour (about 670 Nm^3 H_2 per hour). The lifetime of these systems is in the range of several tens of thousands of hours of operation, which is considered satisfactory for continuous, profitable operation.

Despite the industrial maturity of AECs, they cannot operate at very low-current densities. This limits their flexibility, which is required when coupled with renewable energy sources (e.g., wind and solar energy). Advancements in the materials used for membranes, electrodes, and catalysts are seen as a way to improve this drawback. Asbestos was traditionally used in AECs; however, that is now an illegal material in the majority of the globe. Ceramic/polymer composite materials can replace asbestos, but there is still room for improvement [17]. Additional attempts have also been made to identify alternative electrocatalysts, and some transition metal macrocycles have received attention due to their electrocatalytic properties [18]. Higher efficiencies and performances will be obtainable using such advancements in AEC technologies.

PROTON EXCHANGE MEMBRANE ELECTROLYZER CELL

Research and development of proton exchange membrane electrolyzer cells (PEMECs) have been conducted by many institutions and industries in various countries [18–20], with PEMECs being the probable choice for future utilization of electrolysis using renewable energy sources. Significant progress has been made in PEMEC research in recent decades, with programs like the Japanese WE-NET program, where an energy conversion efficiency of 95.1% was observed and explained by the proximity of the electrolysis cell voltage to the thermoneutral cell voltage (ca. 1.48 V) [21]. PEMECs development with elevated pressures (up to 5.0 MPa) has also shown to be successful [18].

The potential market for PEMECs is similar to AECs; however, the industrially available PEMECs have lower production capacities when compared to the available AECs. The main difference is that AECs are well-established (with decades of technical experience), with an exceptional track record. Moreover, capital expenses are lower, and AECs operate at lower current densities, reducing operational costs. Despite this, PEMECs have a much higher future in terms of room for improvement, and their flexibility with regard to running with low-electrical input favors their utilization with renewable energy networks. In the coming years, it is expected that PEMECs become industrially competitive with AECs with large-scale production of H_2. The main barrier to reach this goal is the required reduction in capital expenses of PEMECs (e.g., a reduction in construction costs of the membrane, bipolar plates, and flow fields). Significant advances in PEMEC technology have been observed in recent years, with room for further improvements. These advancements are the results of many public and industrial research and development programs.

One-way costs can be reduced by bringing down the type and quantity of platinum group metals (PGMs) used. By dispersing nanoparticles on electron-carrier surfaces, this reduction could be met. Similar to what is observed in PEM fuel cell technologies, carbon-supported nanoparticles of platinum are common at the cathode of PEMEC. Mixed oxide (e.g., RuO_2-IrO_2-SnO_2) utilization has also begun to increase at the anode. As a consequence, Pt loadings are reducing significantly, with levels as low as 0.3–0.5 mg per cm^2. The substitution of Pt with Pd (an economically favorable alternative) can also be used as a catalyst and deposited on surfaces [22].

Alternatively, the development of non-PGM electrocatalysts could also lead to reduced costs, and advances have been made recently [23]. Effective results are observed when pure iridium particles are used at the anode in addition to Pt, yielding an efficiency of close to 82% [23]. Polyoxometalates can also be an alternative to Pt; however, their electrochemical performances are lower than observed with the Pt/Ir catalyst discussed earlier. Despite this, the results in efficiency are close to 70% [23]. Cobalt-based clathrochelates [24] are another alternative that may offer exciting perspectives for utilization in the industry. Such approaches are still in the development phases and require further research to realize their potential.

Current density in the PEMEC is also required to be increased, as this will lower capital costs. Additionally, by allowing an increase in operational temperature, the operational cost will be reduced due to a reduction in required cooling. Despite the maximum operating current densities in PEM fuel cells being dependent on the limit of mass transfer (i.e., the transfer of gaseous reactant to the reactive centers), PEMECs do not have such limitations, and significantly high-current densities are possible as a result. However, this creates another problem, as the conductivity of the electrolyte and the ohmic losses become the bottlenecks. Thinner membranes could provide improved efficiencies, but as discussed in Chapter 3, potential gas permeation increases result in reduced faradic efficiencies and reduced gas purities.

The proton exchange membranes require further development in polymer proton conductors allowing operation at elevated temperatures. The applications of PEM fuel cell technologies have driven this research in recent years to improve H_2 transport technologies. Perfluorinated polymer materials, like those produced by Solvay Solexis Co. (Aquivon®), are suggested to be able to operate in temperatures between 120 and 150°C. Also, polybenzimidazole materials that can operate at temperatures around 200°C may offer exciting possibilities for PEMECs in the future. Also, as discussed in Chapter 3, a desirable prospect of PEMECs is that they can increase operating pressures for direct H_2 storage in pressurized vessels; however, this is still a significant research challenge. The central bottleneck is the levels of gas diffusion through the membrane that reduces the current efficiency and leads to explosive H_2/O_2 gas mixtures.

Bipolar plates, flow fields, and current distributors are costly components of PEMECs. This is due to the high raw material costs and the expense of the treatments applied during the engineering stage to reduce contact resistance. By identifying and synthesizing alternative materials, the cost could be reduced in these components shortly. Additional cost-reducing improvements are also possible by increasing the lifetime of the PEMECs (in the range of 10,000–100,000 h lifetime is the current goal). Finally, systems that incorporate both an electrolyzer and a fuel cell (i.e., reversible cells), are exciting in creating stand-alone, decentralized energy systems based on renewable energy sources. As of now, these technologies are still in their primary research phases and will require a significant amount of development to be successfully deployed industrially [25].

SOLID OXIDE ELECTROLYSIS CELLS

High-temperature solid oxide electrolysis cells (SOECs) use ceramic membranes and have been developed by several companies and research centers (e.g., Siemens–Westinghouse, US and Germany). In a SOEC, thin membranes constructed of ceramics are used as the electrolyte, with the electrodes deposited on both surfaces. They can have planar, tubular, and even honeycomb geometries. At high temperatures, there is not a significant requirement for electrocatalysts as in AECs and PEMECs. Therefore, the use of non-PGM catalysts is common; however, it is difficult to eliminate their use as they assure appropriate electrical contacts between the cells. The cathode material is a ceramic-metal alloy containing Ni and Zr. Platinum- or strontium-doped lanthanum manganite is usually the material used at the anode.

SOECs offer the possibility of simultaneous decomposition of water vapor and CO_2. Therefore, as discussed in Chapter 3, SOECs can be used as effective life support systems (e.g., for space and submarine operation). Due to the large heating requirement of SOECs, their advantage becomes clearer when local heat sources (e.g., high-temperature nuclear or concentrated solar power) are available. Despite this, the high operating temperatures required when considering the thin-walled ceramic solid electrolyte films can create problems. One possible route is to develop SOECs that can operate at much lower temperatures (i.e., 500–700°C); however, solid electrolyte materials like Bi_2O_3 have reduced stability at these temperatures, and suffer from mixed conductivity (e.g., both ionic and electronic). These drawbacks could result in a significant reduction in output currents.

With regard to thermodynamics and kinetics, SOECs is an attractive electrolyzer technology as the equilibrium potential difference is less than 1 V at 900°C, and there are negligible diffusion limitations. They do not require significant amounts of precious metal catalysts, and impurities in the feed water do not affect the system significantly. Despite this, structural materials and design of SOECs for high temperature still has some room for improvement, yet the development process for SOECs has reduced in recent years with the focus moving more toward PEMEC technologies.

SOEC components are similar to the technology of their fuel cell alternative; however, although solid oxide cells can be operated as either a fuel cell or an electrolyzer, their operation as an electrolyzer adds some specific challenges concerning the materials used. One such challenge is corrosion (e.g., at the anode due to O_2 evolution) because of the high-operating temperatures. Degradation can be further accelerated by the corrosion of structural components of the cell (e.g., glass or vitro-ceramic seals). Additionally, the feed steam purity has a considerable influence on the rate of degradation [26,27]. Improvements are ongoing about this degradation [18]; however, the exact nature of the degradation mechanisms is still not completely understood [28].

The highly corrosive environment and mechanical and thermal constraints of SOECs are responsible for most performance losses. Lowering the operating temperatures to the 700–800°C range could solve this problem; however, the ionic conductivity of theoxide-ion conductors becomes insufficient. Some scandia-doped zirconium and some nickelates could be used as replacements [18], but further research and development is required in this area.

REFERENCES

[1] Agrafiotis C, Roeb M, Konstandopoulos AG, Nalbandian L, Zaspalis VT, Sattler C, et al. Solar water splitting for hydrogen production with monolithic reactors. Sol Energy 2005;79(4):409–21.

[2] Suman S. Hybrid nuclear-renewable energy systems: a review. J Clean Prod 2018;181:166–77.

[3] Suman S, Khan MK, Pathak M, Singh RN, Chakravartty JK. Rupture behaviour of nuclear fuel cladding during loss-of-coolant accident. Nucl Eng Des 2016;307: 319–27.

[4] Suman S, Khan MK, Pathak M, Singh RN, Chakravartty JK. Hydrogen in Zircaloy: mechanism and its impacts. Int J Hydrogen Energy 2015;40(17): 5976–94.

[5] Augutis J, Martišauskas L, Krikštolaitis R. Energy mix optimization from an energy security perspective. Energy Convers Manag 2015;90:300–14.

[6] Brook BW, Alonso A, Meneley DA, Misak J, Blees T, van Erp JB. Why nuclear energy is sustainable and has to be part of the energy mix. Sustain Mater Technol 2014;1:8–16.

[7] Weimer A, Perkins C, Lichty P, Funke H, Zartman J, Hirsch D, et al. Development of a solar-thermal ZnO/Zn water-splitting thermochemical cycle. Final Report (DE-PS36-03GO93007—Subcontract RF-05-SHGR-006). Colorado, CO, USA, University of Colorado Boulder, 2009

[8] Nastasi B. Hydrogen policy, market, and R&D projects. In: Calise F, D'Accadia MD, Santarelli M, Lanzini A, Ferrero D, editors. Solar Hydrogen Production. Academic Press, 2019, 31–44, doi:10.1016/B978-0-12-814853-2.00002-3.

[9] Muhich CL, Ehrhart BD, Al-Shankiti I, Ward BJ, Musgrave CB, Weimer AW. A review and perspective of efficient hydrogen generation via solar thermal water splitting. Wiley Interdiscip Rev Energy Environ 2016;5(3):261–87.

[10] Perret, R. Solar Thermochemical hydrogen production research (STCH). No. SAND2011-3622. Livermore, CA, United States: Sandia National Lab. (SNL-CA), 2011.

[11] Scheffe JR, Steinfeld A. Oxygen exchange materials for solar thermochemical splitting of H_2O and CO_2: a review. Mater Today 2014;17(7):341–8.

[12] Scheffe JR, Li J, Weimer AW. A spinel ferrite/hercynite water-splitting redox cycle. Int J Hydrogen Ener 2010;35(8):3333–40.

[13] Muhich CL, Evanko BW, Weston KC, Lichty P, Liang X, Martinek J, et al. Efficient generation of H2 by splitting water with an isothermal redox cycle. Science 2013;341(6145):540–2.

[14] Ermanoski I, Miller JE, Allendorf MD. Efficiency maximization in solar-thermochemical fuel production: challenging the concept of isothermal water splitting. Phys Chem Chem Phys 2014;16(18):8418–27.

[15] Kim JT, Johnson A, Miller JE, Stechel EB, Maravelias CT. Fuel production from CO_2 using solar-thermal energy: system level analysis. Energy Environ Sci. 2012;5:8417–29.

[16] Rytter E, Soušková K, Lundgren MK, Ge W, Nannestad ÅD, Venvik HJ, et al. Process concepts to produce syngas for Fischer–Tropsch fuels by solar thermochemical splitting of water and/or CO2. Fuel Process Technol 2016;145:1–8.
[17] Divisek J, Mergel J. Diaphragms for alkaline water electrolysis and method for production of the same as well as utilization thereof. U.S. Patent No. 4,394,244, 1983.
[18] Millet P, Grigoriev S. Water electrolysis technologies. Renew Hydrog Technol Prod Purification, Storage, Appl Saf 2013;19–41.
[19] Millet P, Andolfatto F, Durand R. Design and performance of a solid polymer electrolyte water electrolyzer. Int J Hydrogen Ener 1996;21(2):87–93.
[20] Yamaguchi M, Horiguchi M, Nakanori T, Shinohara T, Nagayama K, Yasuda J. Development of large-scale water electrolyzer using solid polymer electrolyte in WE-NET. In: Proceedings of the 13th World Hydrogen Energy Conference. 2000. p. 274–81.
[21] Hashimoto A, Hashizaki K, Shimizu K. Development of PEM water electrolysis type hydrogen production system for WE-NET. In: Proceedings of the 14th World Hydrogen Energy Conference on CD. 2002.
[22] Grigoriev SA, Millet P, Fateev VN. Evaluation of carbon-supported Pt and Pd nanoparticles for the hydrogen evolution reaction in PEM water electrolysers. J Power Sources 2008;177(2):281–5.
[23] Millet P, Ngameni R, Grigoriev SA, Mbemba N, Brisset F, Ranjbari A, et al. PEM water electrolyzers: from electrocatalysis to stack development. Int J Hydrogen Ener 2010;35(10):5043–52.
[24] Nguyen MTD, Ranjbari A, Catala L, Brisset F, Millet P, Aukauloo A. Implementing molecular catalysts for hydrogen production in proton exchange membrane water electrolysers. Coord Chem Rev 2012;256(21–22):2435–44.
[25] Grigoriev SA, Millet P, Porembsky VI, Fateev VN. Development and preliminary testing of a unitized regenerative fuel cell based on PEM technology. Int J Hydrogen Ener 2011;36(6):4164–8.
[26] Brisse A, Schefold J, Zahid M. High temperature water electrolysis in solid oxide cells. Int J Hydrogen Ener 2008;33(20):5375–82.
[27] Herring JS, O'Brien JE, Stoots CM, Hawkes GL, Hartvigsen JJ, Shahnam M. Progress in high-temperature electrolysis for hydrogen production using planar SOFC technology. Int J Hydrogen Ene 2007;32(4):440–50.
[28] Sohal M. Degradation in solid oxide cells during high temperature electrolysis. No. INL/EXT-09-15617. Idaho National Laboratory (INL), 2009

Chapter | five

Biogas and Hydrogen

Shiplu Sarker[a], Anna S.R. Nordgård[b], Jacob J. Lamb[c], Kristian M. Lien[d]

[a]*Department of Manufacturing and Civil Engineering, ENERSENSE, Faculty of Engineering, NTNU, Trondheim, Norway;* [b]*Department of Biotechnology and Food Science, Faculty of Natural Sciences, NTNU, Trondheim, Norway;* [c]*Department of Electronic Systems, Department of Energy and Process Engineering, ENERSENSE, Faculty of Information Technology and Electrical Engineering, NTNU, Trondheim, Norway;* [d]*Department of Energy and Process Engineering, ENERSENSE, Faculty of Engineering, NTNU, Trondheim, Norway*

CHAPTER OUTLINE

ABSTRACT

Many operational parameters can influence the efficiency of anaerobic digestion to biogas or digestive yield and optimize quality. The most preferred way of maximizing an anaerobic digestion plant is often dependent on how carefully key parameters are selected. The literature on anaerobic digestion optimization is extensive; but given the continuing development and implementation of innovative technology and introduction of increasingly complex systems (i.e., hydrogen injection), the current know-how on hydrogen addition and how this affects the operational range and flexibility in real-life anaerobic digestion system still requires further research. This chapter intends to give an overview of the importance of the effects of hydrogen addition into the anaerobic digestion process.

Hydrogen, Biomass and Bioenergy. http://dx.doi.org/10.1016/B978-0-08-102629-8.00005-0

BIOGAS PRODUCTION PROCESS

Biogas is a high-energy gaseous fuel derived from the anaerobic digestion (AD) process. The raw biogas composing of ~60% CH_4 and ~40% CO_2 can be produced using several solid feedstocks and wastes. With the possibility of renewable energy production being joint to the conversion of solid wastes to biogas is an exciting option as the overall CO_2 emission reduction through this route is higher compared to conventional AD routes. Therefore, increasing efficiency of the existing biogas plants operated with wastes as feedstocks are deemed to be more beneficial and consequently emphasized globally. Furthermore, various pretreatments can be applied to the feedstock to increase the yield of biogas per amount of solid waste [1–4].

The conversion of complex organic matter to methane is done through the cooperation of a wide range of bacteria and archaea. Bacteria convert organic matter into volatile fatty acids (VFAs), single-carbon substrates, CO_2 and H_2 via hydrolysis, acidogenesis, and acetogenesis; while archaea use these substrates to produce methane (Fig. 5.1).

Hydrolysis is the conversion of complex organic matter like proteins, carbohydrates, and fats into smaller molecules (i.e., amino acids, sugars, and fatty acids). This is usually done by extracellular enzymes to facilitate transport into the microbial cell [5]. Acidogenesis (or fermentation)

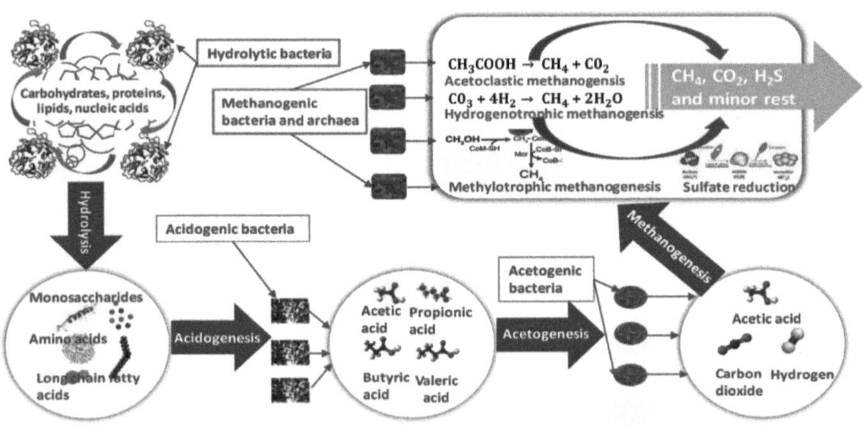

FIGURE 5.1 Biological conversion of complex organic matter to methane (CH_4) and CO_2.

is the conversion of these molecules into VFAs and H_2. VFAs are intermediates in the conversion of organic matter and are essential to monitor [6]. The same bacterial taxa commonly perform hydrolysis and acidogenesis. Several groups of hydrolytic microorganisms are involved in the degradation of several substrate compositions, where the bacteria *Bacteriods*, *Clostridium*, and *Staphylococcus* are significant drivers [7]. Table 5.1 illustrates the different bacterial groups that are involved in the hydrolysis of different substrates.

In the acidogenesis stage, acidogenic bacteria such as *Lactobacillus*, *Streptococcus*, and *Clostridium* transform hydrolysis products (e.g., amino acids and sugars) into VFAs (e.g., acetic acid, butyric acid, and propionic acid), organic acids (e.g., succinic acid and lactic acid), ammonia (NH_3), H_2, carbon dioxide (CO_2), hydrogen sulfide (H_2S), and low-level alcohols [2,7]. The concentration of H_2 produced at this stage affects the final product after digestion and the resulting organic matters such as VFAs are not suited for direct conversion to methane by the methanogens.

Acetogenesis is the bacterial generation of acetate either via the conversion of VFAs by β-oxidation [8] or from CO_2 and H_2 via the Wood-Ljungdahl (WL) pathway [9]. The WL pathway enables acetogenic bacteria to use H_2 as an electron donor, and CO_2 as an electron acceptor and building blocks for biosynthesis. The pathway may be viewed as a series of reactions where two molecules of CO_2 are

TABLE 5.1 Bacterial Groups Involved with the Hydrolysis of Different Substrates [2,5]

Primary Substrate Component	Hydrolyzed Product	Bacterial Groups
Carbohydrates	Soluble sugars	Clostridium, Acitovibrio celluliticus, Staphylococcus, Bacteriodes
Lipids	Higher fatty acids, alcohols, or glycerol	Clostridium, Staphylococcus, Micrococcus
Proteins	Soluble peptides and amino acids	Clostridium, Proteus vulgaris, Peptococcus, Bacteriods, Bacillus, Vibrio

reduced, and four molecules of H_2 are oxidized to produce acetyl-CoA and eventually acetate. The WL pathway produces acetic acid from CO_2 and H_2.

Acetogenic metabolism under methanogenic conditions faces two main challenges—the thermodynamically unfavorable electron disposal, and the energy acquisition from limited energy margins. For example, coupling the reoxidation of NADH with the reduction of H^+ and CO_2 to H_2 and formate is thermodynamically unfavorable. Theoretically, the H_2 gas level must be kept sufficiently low to give a net release of Gibbs free energy. The Gibbs free-energy values given in Table 5.2 illustrate this. In the absence of methanogenic consumption of H_2 during AD, the Gibbs free-energy release for the fermentation of either propionate or butyrate would be positive in value. In this case, these reactions would not proceed. By maintaining the H_2 levels at a low level, these conversions will become energetically favorable (i.e., negative free-energy release). This interplay between H_2-producing and H_2-consuming anaerobes is known as interspecies H_2 transfer (IHT). Without IHT, the overall process of AD would not be possible. The acetogenic bacteria responsible for VFA fermentation and H_2 production are referred to as syntrophs, and they are dependent on IHT with hydrogenotrophic methanogenic archaea. Therefore, they need to stay very close together. Small distances may be obtained by growing in flocs, aggregates, or granules. Syntrophic associations between bacteria and archaea are crucial for VFA conversion.

Methane is produced by methanogenic archaea via three pathways [10,11]—aceticlastic, hydrogenotrophic, and methylotrophic methanogenesis. The majority of the methane is produced by acetotrophic

TABLE 5.2 Fermentation Free-Energy Changes for Standard Versus Typical Reactor Values (kJ/reaction)

Reaction	ΔG^a	ΔG^b
Propionate + $3H_2O \rightarrow$ acetate + H^+ + $3H_2$ + HCO_3^-	+76.2	-5.5
Butyrate + $2H_2O \rightarrow$ 2 acetate + H^+ + H_2	+48.2	-17.6

$^a = \Delta G^0$, *Standard free-energy release with 1M fatty acid and H_2 at 1 atm.*
$^b = \Delta G$, *free-energy release with typical reactor conditions like 1mM fatty acid and H_2 at 10^{-3} atm.*

methanogens via the aceticlastic pathway, which transforms the acetate (resulting from acetogenesis) into CH_4 and CO_2 [12]. In this process, the principle reaction can be interpreted as shown by:

$$CH_3COOH \rightarrow CH_4 + CO_2 \quad (5.1)$$

The hydrogenotrophic group converts H_2 and CO_2 into CH_4 through reactions 5.2 and 5.3 [12]. In this pathway, CO_2 is reduced to methyl-S-CoM via six steps, and the electrons required to reduce methyl-S-CoM to methane are supplied by an external H_2 source (e.g., acetogenic bacteria). From this route, around 30% of the methane may be produced.

$$CO_2 + 4H_2 \rightarrow CH_4 + 2H_2O \quad (5.2)$$

$$4CO + 2H_2O \rightarrow CH_4 + 3CO_2 \quad (5.3)$$

Besides the above two groups, some methane can also be produced by the methylotrophic methanogens [10,11]. Through this pathway, the methyl or trimethylamine component of a given feedstock is transformed into methane by the following chemical:

$$3CH_3OH + 3H_2 \rightarrow 3CH_4 + 3H_2O \quad (5.4)$$

$$4(CH_3)3N + 6H_2O \rightarrow 9CH_4 + 3CO_2 + 4NH_3 \quad (5.5)$$

In methylotrophic methanogenesis, methyl-CoM is produced via the transfer of a methyl group from methanol, methylamines (i.e., mono-, di-, and trimethylamine), or dimethyl sulphide to HS-CoM by enzymatic reactions. Methyl-S-CoM is subsequently disproportionated in a 3:1 ratio; one mole of methyl-S-CoM is oxidized to CO_2 for every three moles of methyl-S-CoM reduced to methane.

COMBINED BIOGAS AND HYDROGEN UTILIZATION

As biogas, the production and utilization of H_2 have been booming in recent years, and H_2 as a fuel is becoming a part of a countless number of applications. With this in mind, processes combining the utilization of biogas and H_2 has gained prominence. Compared to any single process yielding either H_2 or biogas, the merge between the two can result in a significantly positive impact both environmentally and economically. Practically, the synergy between H_2 and biogas can be established through a process generating high methane content gas for subsequent storage or transportation through existing gas grids. This is observed as an advantage due to the reduction of costs, increased progress toward an H_2 society, and lowered environmental impact can be achieved.

By coupling H_2 with biogas, it is also possible to produce various liquid and gaseous fuels ready to be utilized as an alternative to conventional fossil-based fuels or as a complement to these. The process has been implemented already, and many promising results are demonstrated, which favors plant economics and efficiency. Moreover, the biogas and H_2 combination have been examined for separation of the organic component (i.e., CO_2) from the biogas to use as a raw material for many other relevant, innovative approaches. Enhancing efficiency and carbon management of these processes are the critical areas for ongoing research. The development of novel catalysts also contributes to enhancing the efficiency of various processes converting H_2 and biogas to the desired products, which are also important.

Biogas enrichment with hydrogen

Raw biogas produced by AD generally consists of the gas species CH_4, CO_2, H_2S, NH_3, and H_2O (along with the trace amounts of other organic and inorganic components). Methane has a large fraction within the biogas composition with 40%–75%, followed by CO_2 with 25%–55% [13]. Besides AD, biogas can also be collected from landfills with a typical gas composition of 50%–55% CH_4, 37%–45% CO_2, and <1% of other trace compounds [13,14]. Regardless of the production routes, compared to its closest counterpart natural gas (fossil fuel), biogas is energetically inferior due to the high amount of CO_2 and other contaminants within the product [15]. Moreover, the lower heating value of biogas, for example, is roughly 21.5 MJ/Nm^3, while it is around 35.8 MJ/Nm^3 for natural gas [16].

To enhance energy content, the biogas needs to be upgraded [13]. Upgraded biogas can have reduced CO_2 emissions of 75%–200% when compared to that of fossil fuels [17]. Significant reductions in GHG emissions can also be achieved using liquid biofuels derived from upgraded biogas. However, depending upon the applications, upgraded biogas needs to meet the requirements of downstream specifications regarding the levels of contaminants [18]. For instance, high levels of CO_2 within biogas is not desirable in internal combustion engines. A high CO_2 content significantly reduces the energy content of the biogas, therefore, escalating the requirement of gas flow to the combustion engine. Furthermore, the presence of water, H_2S, NH_3, siloxanesn and halocarbons with levels above 1000 ppm tend to cause incomplete combustion and poisonous emissions, making the removal of these components also desirable. Other uses of biogas, such as for gas turbines and micro-turbines for CHP generation require a very low content of water and siloxane (i.e., 0.03–0.1 ppmv), with a tolerable H_2S and halocarbon (Cl-/F-) level 10,000–70,000 ppmv and 200–1500 ppmv, respectively [19].

The prescribed quality of biomethane for natural gas grid injection requires CH_4 concentrations of 80%–96%, CO_2 of 2%–3%, O_2 of 0.2%–0.5%, H_2S of 5 mg/m^3, NH_3 of 3–20 mg/m^3, and siloxanes of 5–10 mg/m^3, respectively [19]. For biogas to reach this quality, various approaches have been used, broadly classified as in-situ and ex-situ upgrading techniques [20,21] (Fig. 5.2). In-situ biogas upgrading involves a liquid-gas phase interaction within the anaerobic reactor moderated in a way that leads to increases in the level of methane within the resulting biogas. By adding certain chemicals (i.e., salts and carbon sources)

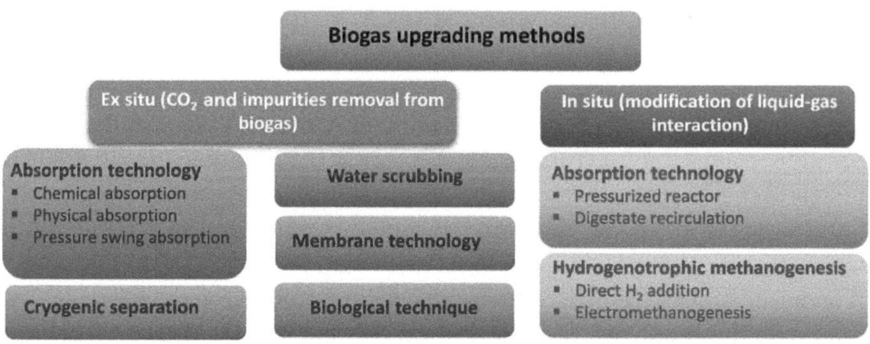

FIGURE 5.2 Different modes of biogas upgrading techniques.

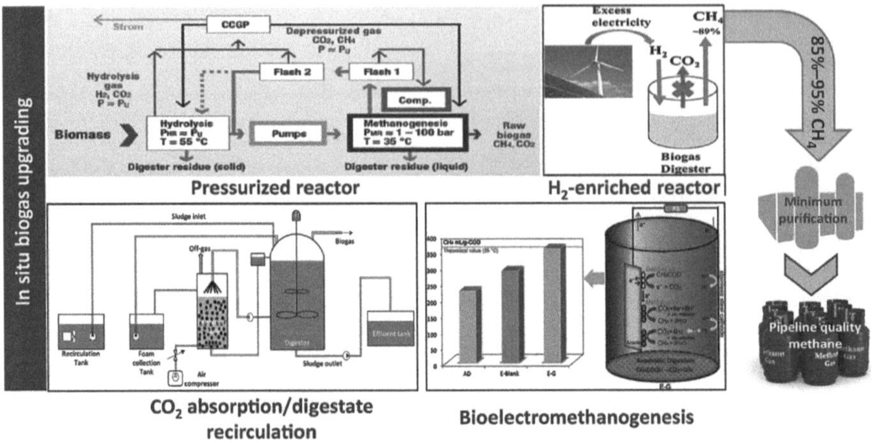

FIGURE 5.3 Biogas upgrading using in-situ techniques.

or gases, or by adjusting some of the process parameters (i.e., pressure and digestate flow), in-situ upgrading can be achieved [22–24]. Different methods of in-situ biogas upgrading techniques are shown in Fig. 5.3.

In-situ hydrogen injection

Traditionally it is believed that 30% of the methane from AD is produced via hydrogenotrophic methanogenesis [25], but adding H_2 directly to an anaerobic reactor may change the microbial community composition, promoting the hydrogenotrophic methanogenesis pathway and consequently enhancing the biological conversion of CO_2 into methane. Utilizing this concept (known as biomethanation), an increase of CH_4 yields up to 20%–40%, and a possibility of up to 90% has been observed [25–29]. It has generally been believed that aceticlastic methanogenesis has been the dominant pathway in most biogas reactors producing approximately 70% of the collected methane. A recent study conducted in Sweden using samples from nine randomly selected large-scale digesters revealed that methane was generated through syntrophic acetate oxidation (SAO) coupled with hydrogenotrophic methanogenesis, rather than through aceticlastic methanogenesis [9]. Thus, hydrogenotrophic methanogenesis appears to be of much greater importance for the biogas process than assumed previously, and maybe aceticlastic methanogenesis should no longer be considered to be the main pathway [9].

The significant advantage of the in-situ technique is that it allows existing biogas plants to be utilized for H_2 addition and available natural gas infrastructure for upgraded biomethane injection, therefore, eliminating the need for H_2 storage [30]. However, for this technology to become cost-effective, a reliable and cheap supply of H_2 would be necessary. H_2 can be obtained from various supply channels; however, electrolysis using surplus electricity from wind and solar sources might suit perfectly to specific contexts where electricity is less expensive, such as in Norway.

Alternatively, current research has identified a set of bacterial species that when subjected to electrical potential are capable of producing endogenous H_2 as a source to promote hydrogenotrophic methanogenesis. The concept relies on interdisciplinary research with the application of this technology only limited to lab-scale studies at this time [26,27]. This is mainly due to its low volumetric CH_4 production rates [31] and the technical challenges associated with the optimization of the process [24]. For example, an H_2 injection exceeding a 4:1 stoichiometric ratio between H_2 and CO_2 will tend to deplete CO_2, resulting in the rise of pH [26]. This will inhibit the autotrophic hydrogenotrophic methanogenesis as this requires CO_2 as a carbon source to reduce H_2 into CH_4 [32]. Moreover, H_2 dissolves very poorly in the aqueous phase [33], and the extent hydrogenotrophic methanogens convert H_2 into CH_4 strongly depends on the efficiency at which H_2 dissolves into liquid phases. The H_2-to-liquid interaction can be expressed as [20]:

$$r_t = 22.4 k_L a (H_{2gTh} - H_{2l}) \tag{5.6}$$

where,

r_t: H_2 liquid mass transfer rate (L/(L.d)

22.4: gas volume to mole ratio (1 mol gas corresponds to 22.4 L gas at STP)

$k_L a$: gas transfer coefficient (per day)

H_{2gTh}: H_2 concentration in the gas phase (mol/L)

H_{2l}: H_2 dissolved in the liquid phase (mol/L)

Based on this, the theoretical improvement in H_2 liquid mass transfer can be achieved by improving $k_L a$ using techniques like modulated mixing speed [30,34], gas recirculation [35], changing diffusion devices [26,36], adding packing materials (increasing gas-liquid mass transfer) [20], a modified reactor design using a trickle bed [21], and the use of an up-flow anaerobic sludge blanket (UASB) reactor [20]. These approaches have so far been applied in lab-scale studies; however, to drive the in-situ biogas upgradation technology toward commercial development, full-scale studies will be urgently necessary. The H_2 addition to an AD once commercialized successfully can open a new door to methane upgrading from biogas plants.

Ex-situ hydrogen injection

Unlike the previous concept where H_2 is added to a live AD process, in the power to gas (PtG) concept,[a] the H_2 can be added to the biogas after being produced from the AD process. Afterward, the mixture of H_2 and biogas can be dealt with in an external reactor facilitating catalytical/biological conversion to high-quality biomethane. PtG can link the electricity and gas grids by using surplus renewable electricity for H_2 production. The further biogas conversion to storable biomethane for direct injection into existing natural gas grid results in benefits such as fossil fuel-free transportation, improved life cycle emissions from the power generation plants and vehicles, and aides the transition to the future H_2 society. In a PtG system using renewable energy and biogas, the following two steps are used:

1. Electrolysis; and,
2. Methanation.

In electrolysis, H_2 production is achieved using renewable electricity. There are various electrolysis options (described in Chapter 3), which can be directly applied to PtG. The most common techniques are water electrolysis through AECs (alkaline electrolysis cells), PEMECs (proton exchange membrane electrolysis cells), and SOECs (solid oxide electrolysis cells).

Among the different water electrolysis techniques, AEC is well known, and most considered in PtG. AECs offer both atmospheric and high-pressure operation modes. The high-pressure operation mode enables the production of compressed H_2 suiting direct grid injection.

---

[a] Power to gas concept can be applied through different pathways for a range of applications including PtG to electrolyzer, power to power, power to microgrid, power to renewable natural gas to pipeline/storage, power to renewable fraction of petroleum fuels, power to zero emission transportation and many more.

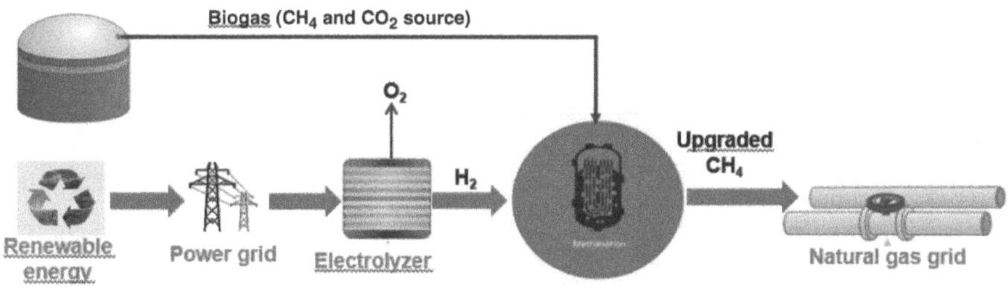

FIGURE 5.4 Power to H_2 to natural gas end users.

This reduces additional cost and energy input to PtG plants. A typical layout of the PtG process, considering catalytic and biological methanation, is shown in Fig. 5.4.

BIOGAS FUEL STORAGE

Four types of pressure vessels can be used for storing pressurized gas [37].

1. Fully metallic pressure vessels (contain pressures up to 50 MPa);
2. Steel pressure vessel with a glass fiber composite overwrap (high-pressure tolerance);
3. Full composite wrap with metal liner (reliable for 45 MPa working pressures); and,
4. Fully composite (can withstand pressures up to 100 MPa).

Ideally, biogas would be upgraded and injected directly into the natural gas infrastructure. When considering the required increases in purity needed to be performed after AD, it becomes clear that liquid storage is more favorable for high-quality biogas. Heat exchangers are among the principal components of gas liquefaction systems. To achieve excellent overall energy performance of the system, cryogenic heat exchangers should be designed to operate at the lowest possible temperature differences. This requires large heat transfer areas and bulky equipment with high initial costs [38,39].

The heat transfer occurs in a variety of process applications [40], and these can be summarized into three categories [41]:

1. Process cooling;
2. Separation of gas mixtures; and,

3. Liquefaction for transportation and storage.
The liquefied biogas (LBG) technology provides an economically feasible way of transporting biogas long distances. Biogas liquefaction requires large cryogenic heat exchangers, where biogas is cooled, liquefied, and subcooled to 111 K. Some liquefaction cycles utilize separate heat exchangers to perform the precooling or subcooling duty, while others may require only a single multi-stream heat exchanger per production train, commonly referred to as the main heat exchanger [42]. The economic feasibility of many LBG projects, considering the associated capital and operating costs, is, to a large extent, controlled by the performance of the main cryogenic heat exchanger [42].

Liquefaction shrinks the volume of the gas almost 600 times, making it easier and safer to store and transport. Once delivered, it can be converted back into gas at regasification facilities, usually involving large heat exchangers again.

REFERENCES

[1] Lamb JJ, Islam MH, Hjelme DR, Pollet BG, Lien KM. Effect of power ultrasound and Fenton reagents on the biomethane potential from steam-exploded birchwood. Ultrason Sonochem 2019;58:104675. doi:10.1016/j.ultsonch.2019.104675.

[2] Sarker S, Lamb JJ, Hjelme DR, Lien KM. A Review of the role of critical parameters in the design and operation of biogas production plants. Appl Sci 2019;9(9): 1915.

[3] Lamb JJ, Hjelme DR, Lien KM. Carbohydrate yield and biomethane potential from enzymatically hydrolysed saccharina latissima and its industrial potential. Adv Microbiol 2019;9(4):359–71.

[4] Lamb JJ, Sarker S, Hjelme DR, Lien KM. Fermentative bioethanol production using enzymatically hydrolysed *Saccharina latissima*. Adv Microbiol 2018;8(5): 378–89.

[5] Angelidaki I, Karakashev D, Batstone DJ, Plugge CM, Stams AJM. Biomethanation and its potential in: methods in enzymology. In: Amy CR, Stephen WR, editors. San Diego, USA: Academic Press; 2011.

[6] Lamb JJ, Bernard O, Sarker S, Lien KM, Hjelme DR. Perspectives of optical colourimetric sensors for anaerobic digestion. Renew Sustain Energy Rev 2019;111: 87–96.

[7] Nayono SE. Anaerobic digestion of organic solid waste for energy production. Karlsruhe, Germany: KIT Scientific Publishing; 2009.

[8] Schink B. Energetics of syntrophic cooperation in methanogenic degradation. Microbiol Mol Biol Rev 1997;61(2):262–80.
[9] Müller B, Sun L, Schnürer A. First insights into the syntrophic acetate-oxidizing bacteria–a genetic study. Microbiol Open 2013;2(1):35–53.
[10] Garcia J-L, Patel BKC, Ollivier B. Taxonomic, phylogenetic, and ecological diversity of methanogenic archaea. Anaerobe 2000;6(4):205–26.
[11] Gerardi M.H. The microbiology of anaerobic digesters, 1st ed. Somerset NJ: Wiley, 2003.
[12] André L, Ndiaye M, Pernier M, Lespinard O, Pauss A, Lamy E, et al. Methane production improvement by modulation of solid phase immersion in dry batch anaerobic digestion process: dynamic of methanogen populations. Bioresour Technol 2016;207:353–60.
[13] Kadam R, Panwar NL. Recent advancement in biogas enrichment and its applications. Renew Sustain Energy Rev 2017;73:892–903.
[14] Eklund B, Anderson EP, Walker BL, Burrows DB. Characterization of landfill gas composition at the fresh kills municipal solid-waste landfill. Environ Sci Technol 1998;32(15):2233–7.
[15] Sarker S, Lamb JJ, Hjelme DR, Lien KM. Overview of recent progress towards in-situ biogas upgradation techniques. Fuel 2018;226:686–97. Available from: https://doi.org/10.1016/j.fuel.2018.04.021.
[16] Tarang B, Nikhil T, Gaurav C. Upgradation of biogas using combined method of alkaline water scrubbing and adsoption through carbon molecular sieve. Int J ChemTech Res 2013;5(2):886–90.
[17] Papacz W. Biogas as vehicle fuel. J KONES 2011;18:403–10.
[18] Singhal S, Agarwal S, Arora S, Sharma P, Singhal N. Upgrading techniques for transformation of biogas to bio-CNG: a review. Int J Energy Res 2017;41(12):1657–69.
[19] Soreanu G, Beland M, Falletta P, Edmonson K, Svoboda L, Al-Jamal M, et al. Approaches concerning siloxane removal from biogas–a review. Can Biosyst Eng 2011;53(8):1–18.
[20] Bassani I, Kougias PG, Angelidaki I. In-situ biogas upgrading in thermophilic granular UASB reactor: key factors affecting the hydrogen mass transfer rate. Bioresour Technol 2016;221:485–91.
[21] Rachbauer L, Voitl G, Bochmann G, Fuchs W. Biological biogas upgrading capacity of a hydrogenotrophic community in a trickle-bed reactor. Appl Energy 2016;180:483–90.
[22] Hayes TD, Isaacson HR, Pfeffer JT, Liu YM. In situ methane enrichment in anaerobic digestion. Biotechnol Bioeng 1990;35(1):73–86.
[23] Lemmer A, Chen Y, Wonneberger A-M, Graf F, Reimert R. Integration of a water scrubbing technique and two-stage pressurized anaerobic digestion in one process. Energies 2015;8(3):2048–65.
[24] Mulat DG, Mosbæk F, Ward AJ, Polag D, Greule M, Keppler F, et al. Exogenous addition of H2 for an in situ biogas upgrading through biological reduction of carbon dioxide into methane. Waste Manag 2017;68:146–56.

[25] Bassani I, Kougias PG, Treu L, Angelidaki I. Biogas upgrading via hydrogenotrophic methanogenesis in two-stage continuous stirred tank reactors at mesophilic and thermophilic conditions. Environ Sci Technol 2015;49(20):12585–93.
[26] Luo G, Angelidaki I. Co-digestion of manure and whey for in situ biogas upgrading by the addition of H 2: process performance and microbial insights. Appl Microbiol Biotechnol 2013;97(3):1373–81.
[27] Luo G, Johansson S, Boe K, Xie L, Zhou Q, Angelidaki I. Simultaneous hydrogen utilization and in situ biogas upgrading in an anaerobic reactor. Biotechnol Bioeng 2012;109(4):1088–94.
[28] Luo G, Angelidaki I. Hollow fiber membrane based H 2 diffusion for efficient in situ biogas upgrading in an anaerobic reactor. Appl Microbiol Biotechnol 2013;97(8):3739–44.
[29] Martin MR, Fornero JJ, Stark R, Mets L, Angenent LT. A single-culture bioprocess of *Methanothermobacter thermautotrophicus* to upgrade digester biogas by CO_2-to-CH_4 conversion with H_2. Archaea 2013;11. 2013. doi:10.1155/2013/157529.
[30] Luo G, Angelidaki I. Integrated biogas upgrading and hydrogen utilization in an anaerobic reactor containing enriched hydrogenotrophic methanogenic culture. Biotechnol Bioeng 2012;109(11):2729–36.
[31] Rittmann SKMR. A critical assessment of microbiological biogas to biomethane upgrading systems. In: Guebitz G, Bauer A, Bochmann G, Gronauer A, Weiss S, editors. Biogas science and technology. Advances in biochemical engineering/biotechnology, vol. 151. Cham: Springer, 2015.
[32] Agneessens LM, Ottosen LDM, Voigt NV, Nielsen JL, de Jonge N, Fischer CH, et al. In-situ biogas upgrading with pulse H2 additions: the relevance of methanogen adaption and inorganic carbon level. Bioresour Technol 2017;233:256–63.
[33] Pauss A, Andre G, Perrier M, Guiot SR. Liquid-to-gas mass transfer in anaerobic processes: inevitable transfer limitations of methane and hydrogen in the biomethanation process. Appl Environ Microbiol 1990;56(6):1636–44.
[34] Bhattacharyya D, Singh KS. Understanding the mixing pattern in an anaerobic expanded granular sludge bed reactor: effect of liquid recirculation. J Environ Eng 2009;136(6):576–84.
[35] Guiot SR, Cimpoia R, Carayon G. Potential of wastewater-treating anaerobic granules for biomethanation of synthesis gas. Environ Sci Technol 2011;45(5):2006–12.
[36] Díaz I, Pérez C, Alfaro N, Fdz-Polanco F. A feasibility study on the bioconversion of CO_2 and H_2 to biomethane by gas sparging through polymeric membranes. Bioresour Technol 2015;185:246–53.
[37] Barthelemy H, Weber M, Barbier F. Hydrogen storage: recent improvements and industrial perspectives. Int J Hydrogen Energy 2017;42(11):7254–62.
[38] Pacio JC. Multiscale thermo-hydraulic modeling of cryogenic heat exchangers, PhD thesis, Faculty of Science and Engineering, Department of Energy and Process Engineering, Norwegian University of Science and Technology, 2012. Available from: https://ntnuopen.ntnu.no/ntnu-xmlui/handle/11250/234575.
[39] Shah RK, Sekulic DP. Fundamentals of heat exchanger design. Hoboken, NJ: John Wiley & Sons; 2003.

[40] Barron RF. Cryogenic heat transfer. Ann Arbor, MI: Edwards Brothers; 1999.
[41] Fikiin K, Stankov B, Evans J, Maidment G, Foster A, Brown T, et al. Refrigerated warehouses as intelligent hubs to integrate renewable energy in industrial food refrigeration and to enhance power grid sustainability. Trends Food Sci Technol 2017;60:96–103.
[42] Popov D, Fikiin K, Stankov B, Alvarez G, Youbi-Idrissi M, Damas A, et al. Cryogenic heat exchangers for process cooling and renewable energy storage: a review. Appl Therm Eng 2019;153:275–90. doi:10.1016/j.applthermaleng.2019.02.106.

Chapter | six

Thermochemical Production of Fuels

Erling Rytter[a], Magne Hillestad[a], Bjørn Austbø[b], Jacob J. Lamb[c], Shiplu Sarker[d]

[a]*Department of Chemical Engineering, ENERSENSE, Faculty of Natural Sciences, NTNU, Trondheim, Norway;* [b]*Department of Energy and Process Engineering, ENERSENSE, Faculty of Engineering, NTNU, Trondheim, Norway;* [c]*Department of Electronic Systems, Department of Energy and Process Engineering, ENERSENSE, Faculty of Information Technology and Electrical Engineering, NTNU, Trondheim, Norway;* [d]*Department of Manufacturing and Civil Engineering, ENERSENSE, Faculty of Engineering, NTNU, Trondheim, Norway*

CHAPTER OUTLINE

ABSTRACT

Thermochemical production of fuels is based on the use of biomass to develop useful fuels. Biomass-derived fuels seem to be a promising path in reducing our societies' dependance on fossil fuels, and various routes are available for biomass processing. Examples include hydrotreatment, hydrothermal liquefaction, hydroprocessing, hydrogenation, reforming, and gasification. This chapter gives an overview of these thermochemical productions processes. Also, this chapter provides scope for the implementation of such processes within research and industry.

Hydrogen, Biomass and Bioenergy. http://dx.doi.org/10.1016/B978-0-08-102629-8.00006-2

INTRODUCTION

Production of fuels from biomass follows a few main paths or process types that are widely different. It is customary to classify biofuel into standard, or 1st generation, and advanced, or 2nd generation. First generation biofuels are made from vegetable oil or sugar that requires only conventional technology for upgrading. Second generation biofuel is made from lignocellulosic biomass like agricultural waste, forest residues, and municipal waste, or dedicated energy crops [1]. Characteristics of 2nd generation biofuel are high utilization of the biomass and very low greenhouse gas (GHG) impact. Still, the amount of produced fuel is low on a weight-to-weight basis—unless H_2 is added [2].

Depending on how severe the biomass molecular constituents are modified or broken down, the processes described here are classified as follows:

1. Hydrotreated vegetable oil (HVO). Hydrogenation of natural bio-oil is the least severe treatment, and the product is commercialized for diesel vehicles. The acronym HVO includes for convenience also products made from animal fat and wax as these have equivalent properties;
2. Hydrothermal liquefaction (HTL) of biomass is a catalytic, thermochemical process that breaks down cellulose and hemicellulose to smaller molecules for further separation and processing. Therefore, it is also called catalytic depolymerization or direct liquefaction. This is the basis of a biorefinery for making a multitude of chemicals and fuels;
3. Direct hydro processing resembles HTL but is operated with co-feeding H_2 at a higher temperature and lower pressure than HTL (i.e., more typical for an oil refinery process); and,
4. Biomass to liquids (BtL) is multistep chemical processing where biomass first is gasified to synthesis gas (syngas) containing carbon monoxide, H_2, and CO_2, before polymerization to liquid hydrocarbon fuels. Methanol can also be produced from syngas and is often included as a BtL product.

There are also further 1st generation processes for fuel generation that have its origin in biomass as feedstock. Most significant is biofuels made by transesterification of vegetable oil (called FAME; Fatty Acid Methyl

Ester), that presently is a primary blending component in diesel. Some FAME types have been challenged due to extensive land use, deforestation, and questionable impact on GHG reduction. This is a general concern for most biofuels: What is the impact on global climate, and what are the local NO_x and particle emissions when used in a combustion engine? Another fuel-blending product that is particularly important in Brazil is ethanol made by fermentation of sugar canes. However, these products do not depend on reaction with H_2 and are not considered further other than as comparative examples.

Fermentation and gasification inevitably give rise to large amounts of CO_2, let us call this bio-CO_2. Reacting bio-CO_2 at high temperature with H_2 again gives syngas for further fuel synthesis. Thus, a fifth category can be added:

5. Hydrogenation of bio-CO_2 to synthesis gas. The product obtained after polymerization sometimes is called blue crude.

All five bulleted process types depend on reacting biomass, or broken-down biomass, with H_2 [2]. Use of H_2 in upgrading raw biomass follows naturally because biomass is rich in carbon and oxygen and short in H_2 compared to the final product, fuels [2]. Another way of expressing this is that biomass is partly combusted hydrocarbons and needs energy (i.e., H_2) to be lifted to the energy density of standard gasoline or diesel. Liquid conventional fuels have the unique properties of high-energy density, and ease of transportation and storage.

Hydrotreated vegetable oil (HVO)

HVO is a renewable type of diesel produced by adding H_2 to vegetable oil. The product has excellent properties with reduced emissions, and it can be used in all types of diesel engines as a blending component or pure form. HVO is in some respects similar to FAME as both use vegetable oil as raw material.

HVO PRODUCTION AND SUPPLY

Vegetable and animal fats consist mainly of triglyceride. This is an ester composed of fatty acid, that is, an organic acid with a long paraffinic backbone (R), and glycerol. R is a straight chain hydrocarbon typically in the C_{15}–C_{18} carbon range that, depending on the feedstock, is partly unsaturated. Fig. 6.1 compares how triglyceride is hydrogenated

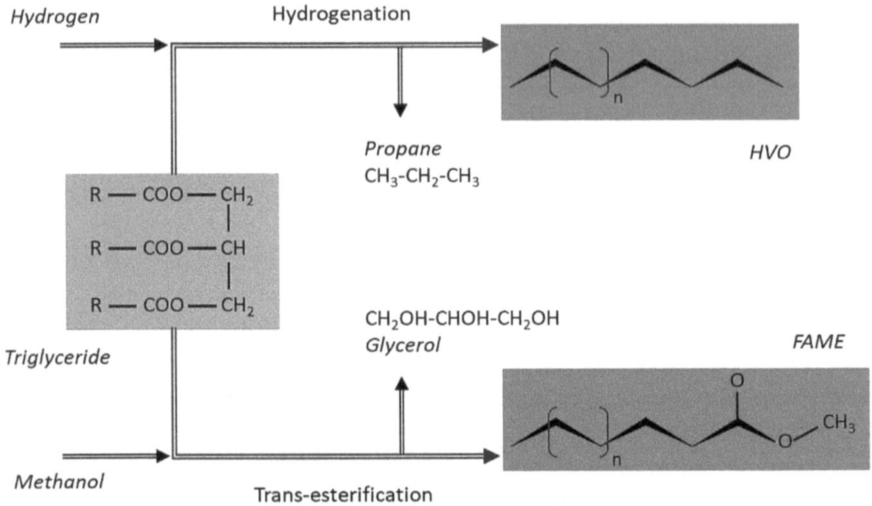

FIGURE 6.1 Synthesis of HVO compared with FAME.

to HVO and, alternatively, how methanol splits of glycerol to make the conventional biofuel FAME. The mechanisms are conceivably simple. In both cases, the bonds between the ester groups and the C_3 central backbone of triglyceride is broken. A significant difference is that HVO is fully hydrogenated, meaning that all oxygen atoms and smaller amounts of sulfur and nitrogen are removed. The resulting paraffin is excellent fuel but needs to be isomerized for optimal properties, particularly reduction of freezing point. FAME maintains the ester groups during the synthesis, now with methyl from methanol at the end. The processing conditions are also very different. Hydrogenation requires a catalyst, high pressure (30–70 bar), and high temperature (300–350°C). Transesterification takes place at much milder conditions; at ambient pressure, 50–80°C and in a basic solution. Propane is a useful by-product from HVO production, while a large amount of glycerol has low value.

Raps oil, palm oil, animal fat, fish oil, and used frying oil or certain industrial oils are among conventional oils that can favorably be hydrotreated. Neste Oil of Finland is the world's largest manufacturer of HVO. Other European producers are Preem of Sweden and ENI of

Italy. Neste Oil is producing mainly from wastes in the fish and animal farm industries, with a supplement of palm oil. They make pure HVO in a dedicated part of their refinery, whereas Preem blend sulfate cellulose oil with conventional crude oil before processing in the hydrotreater. The price of FAME from the refinery has the last five years been around 0.7 USD per L, and the price of HVO is assumed to be somewhat higher.

As the by far largest producer worldwide, Neste produces 2.6 million tons of HVO (2017), more than the total diesel consumption in Norway. They use a process called NExBTL [3]. The collected vegetable and animal oils are first pretreated by aqueous acid and caustic wash before hydrotreating, hydroisomerization, hydrocracking, and distillation. The following companies also have developed similar HVO production technologies: IFP Axens (Vegan); Honeywell UOP (Green Diesel), Haldor Topsoe (Hydroflex), and ENI (Ecofining).

HVO FUEL PROPERTIES

In contrast to FAME, there are no issues for HVO with microbial growth or blending with water. Properly hydroisomerized, cold weather properties are excellent with no problems with filter clogging. Hydrotreating removes double bonds, thereby securing excellent storage stability with little chance of forming gums. The European diesel standard (EN590) restricts conventional biodiesel content to maximum 7%, and many engine manufacturers do not recommend the use of conventional FAME-type biodiesel at all in fear of possible engine damage. The EN590 standard specifies the density of winter diesel to be above 0.80 kg/L. This density limitation restricts blending of HVO that typically has a density of 0.78 kg/L. The consequence of slightly lower density is that, with a given fuel pump, the power of the engine is marginally reduced due to a lower energy content per volume; but it is higher per weight unit. Energy contents (lower heating value) are compared in Table 6.1. However, these are exempt from the density specification for use in dedicated fleets of vehicles. The ASTM D976 standard in the USA and CAN/CGSB 3.517 standard in Canada allow the use of 100% HVO. In contrast to fossil diesel, HVO is paraffinic and therefore is comparatively clean burning. It is fully miscible with fossil diesel.

TABLE 6.1 Energy Contents in Biofuel and Fossil Diesel

Fuel	Density (g/L)	Volumetric Heating Value (MJ/L)	Weight-Based Heating Value (MJ/kg)
Fossil diesel	0.835	36.0	43.1
HVO	0.78	34.4	44.1
FAME	0.88	32.7	37.2

There is a specific European standard EN15940 for HVO that also covers BtL diesel, see section Direct hydroprocessing (later in this Chapter) as these are very close in chemical structure and composition. Some properties and specifications are summarized in Table 6.2. As long as the specifications are met, there is no requirement to label pumps with the type of diesel. In addition to density, aromatics content, and blending with conventional biodiesel mentioned earlier, a few other properties are worth mentioning. Both HVO and BtL diesel are bright and clear with an appearance resembling water. Cetane number describes compressibility

TABLE 6.2 Typical Properties of HVO and BtL Diesel Compared with Standards

Property	Unit	HVO/BtL	EN15940	EN590	ASTM: D975
Appearance		Clear			
Cetane number		>70	≥70	≥51	≥40
Density (15°C)	kg/L	0.77–0.79	0.765–0.800	≥0.800	—
Aromatics	%(w/w)	<0.5	≤1.1	—	≤35
Sulfur	ppm	<1	≤5	≤10	≤15
FAME	%(v/v)	0	≤7	≤7	—
Flash point	°C	>61	>55	>55	>52

TABLE 6.3 Diesel Fuel Tax Levels in Finland (Euro Cents Per Liter)

Fuel	Tax	Reduction (%)
Fossil fuel	50.6	—
HVO from palm-oil	34.0	33
HVO from waste	25.2	50
FAME from palm-oil	37.8	25

before self-ignition. It is an inverse function of the fuel's ignition delay; that is, time from compression start and a noticeable pressure increase. Due to the very high cetane number, it follows that HVO can be blended with fossil-type diesel that does not comply with the specifications, typically high aromatic diesel. Thereby, low-value gas oil that otherwise would have been used as fuel oil now can be sold as vehicle fuel. BtL and HVO are virtually free of sulfur; HVO due to extensive hydrotreating and BtL simply because sulfur has to be removed to ppb levels not to deactivate the catalyst. The flash point is a measure of volatile material above the fuel and the lowest temperature whereby it can be ignited with an ignition source. All in all, it is evident that HVO and BtL diesel have superior properties and can be used neat or to blend with other fuels to improve their properties.

LOCAL EMISSIONS AND GREENHOUSE GASES

Many countries offer a reduction in fuel tax based on a reduction in local and GHG emissions. An example from Finland is given in Table 6.3. The reduction is lower when produced from vegetable oil like palm oil than from biomass-based waste. [a]Why the maximum reduction is not larger than 50% is due to a general energy component in the tax reflecting local emissions, typically NO_X. To be qualified for specific

[a] In January 2018, the European parliament voted in favor of exempting palm oil from being a viable source for biofuel. Further, all biofuels based on using agricultural land are to be gradually reduced. Simultaneously, biofuel blending is to be increased. If this goes further to a directive, it means a boost for biofuels based on lignocellulose [17].

tax reductions depends on fulfilling standard criteria of the European Union as to the effect on greenhouse gas emissions. To calculate GHG effects of fuels is a science of its own where numerous factors have to be taken into account. The most complicated is probably the alternative land use component. To what extent does the bio-oil production reduce food production?

Moreover, is biofuel production based on deforesting? How much methane is released from the soil during ploughing? Further, there is an effect of the biomass growth cycle. How long time does it take for the biomass to grow to reach original volume after harvesting? This is particularly important for woody biomass that takes decades to grow. Add to this the energy source for production, transportation, and fertilizers, and it becomes a complex equation.

Several well-to-wheel calculations of fuels have been reported, and in EU, there are standardized methods of doing this [4]. It is not pertinent to go into details here, but the compilation made by Neste based on testing of three different vehicle fleets is revealing, as shown in Fig. 6.2 [5]. The reduction in equivalent CO_2 emission is significant for all biofuels, with HVO being slightly better than FAME. E85, that

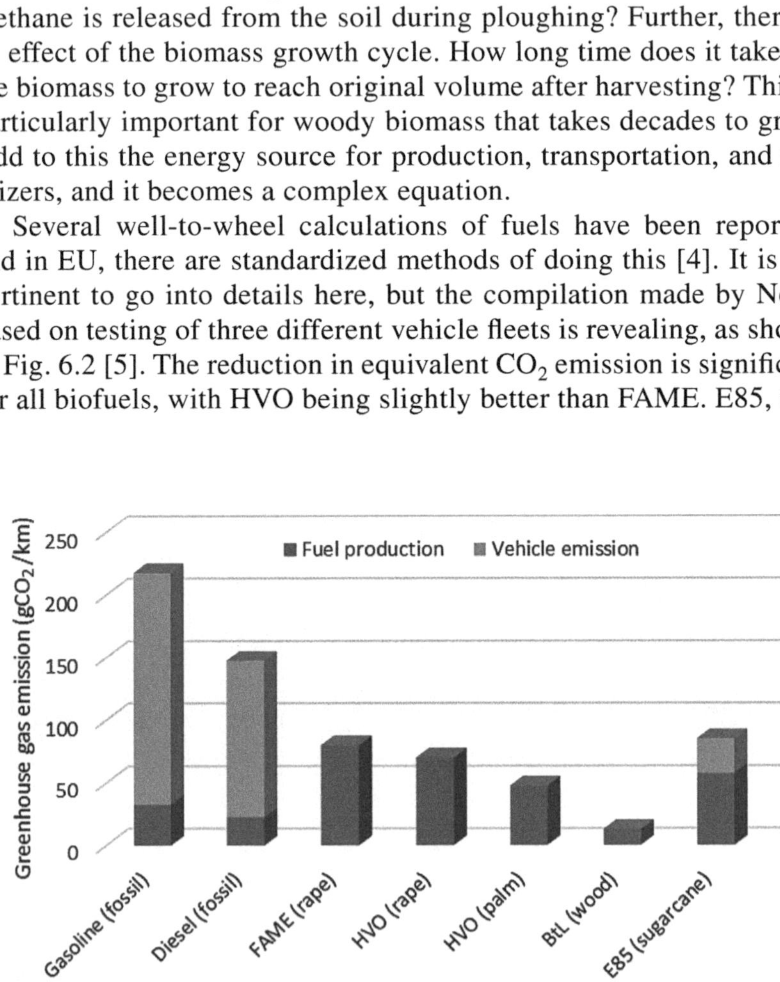

FIGURE 6.2 Equivalent CO_2 well-to-wheel emissions of a mid-size class car for different fuels. *Data from Ref.* [5].

is, gasoline blended with 85% ethanol from sugarcanes, is hampered by low-energy content and emissions from the residual gasoline. Note that the best fuel by far is based on gasification of wood and the BtL technology.

Measurements of tailpipe emissions compared to fossil diesel also show significant benefits. Apart from no sulfur (SO_2), the most substantial response is for CO and hydrocarbons with a typical reduction in the 30%–40% range. Particulates that together with sulfur and NO_x are particularly important for the urban atmosphere is typically reduced by 20%. The reduction is valid for all particle sizes. There is, however, only a marginal reduction in NO_x emission. We conclude that HVO, and especially BtL diesel, is superior regarding environmental impact. It is not to be forgotten. However, this production method of biodiesel is costly and energy consumption compared to regular diesel. It takes about six times as much energy to produce 1 L of fuel from rape seeds or palm oil compared to upgrading crude oil in a refinery. This energy inefficiency does not appear explicitly in the GHG compilation as biomass is regarded as CO_2 neutral.

Hydrothermal liquefaction

Hydrothermal liquefaction (HTL) differs from the other technologies for fuel production from biomass under the influence of H_2 in that the H_2 comes from water directly (i.e., without first splitting water by solar energy or electrolysis). It is also called direct liquefaction as it does not involve gasification. HTL mimics the natural geological process of thermochemical conversion of biomass to petroleum at an accelerated speed. Hydrothermal liquefaction is thermal depolymerization of wet biomass under moderate temperature and high pressure, often using a homogeneous or heterogeneous catalyst. The product is bio-oil; also called biocrude. The bio-oil can be used as fuel in heavy-duty engines but requires typically further processing and separation in a biorefinery to several useful products, including biofuel.

A significant research effort is put into this technology internationally, and there has been a limited number of industrial demonstrations. Most laboratory tests are performed in batch reactors, so-called one-pot tests, and there is only a limited number of continuous flow investigations. The latter is most relevant for an industrial process. The principle of using hot, reactive water to convert biomass was found as early as in the 1020s and was further developed under high-pressure conditions at PERC (Pittsburg Energy Research Center) in the 1970s [6]. A 100 kg/h demonstration plant was erected at the Albany Biomass Liquefaction Experimental Facility in Oregon, US. All kinds of biomass are considered for HTL, including waste, crop residues, manure, and algae.

HTL CHEMISTRY

HTL is operated at temperatures between 250 and 400°C in the pressure range 50–250 bar with residence times from several minutes to a few hours [7,8]. At these conditions, water is sub-critical and approaching critical conditions (Fig. 6.3).

The H_2 atoms become very reactive and water acts to a large degree as a solvent for the biomass. Main reactions that proceed are cracking of long-chained hydrocarbons and removing oxygen by dehydration making H_2O and decarboxylation making CO_2 (Fig. 6.4). The reaction network and mechanisms, however, are complex and have been suggested to involve a liberated proton; H.

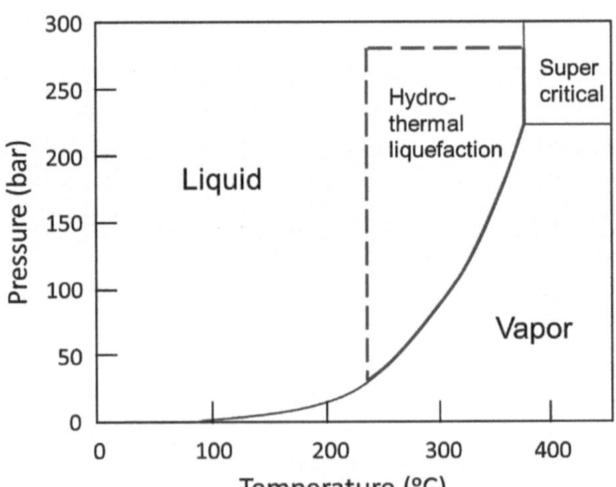

FIGURE 6.3 Phase diagram of water with hydrothermal region.

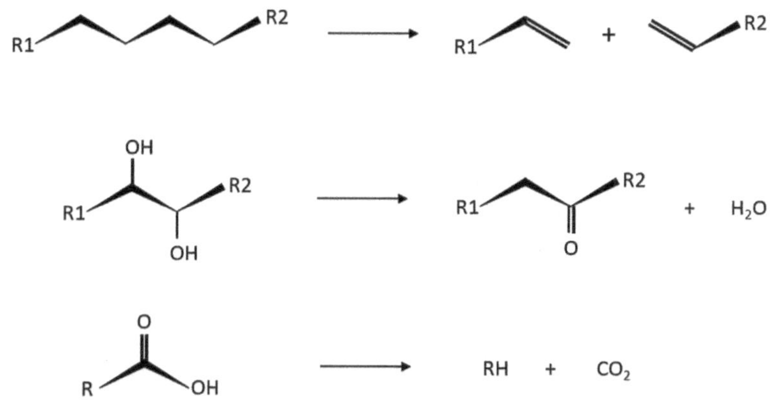

FIGURE 6.4 Typical reactions in hydrothermal liquefaction. From top: cracking; dehydration to a ketone; decarboxylation. (Additional hydrogen atoms not shown.)

A large number of different catalysts and catalyst formulations have been tested. They span from alkali like KOH or $Ca(OH)_2$; carbonates as Na_2CO_3, Ni, or Co transition metals; and the noble metals Pt, Rh, and Ru [7]. Both protein, lipid, cellulose, and hemicellulose can be converted; unfortunately, the aromatic ring structures of lignin are not affected [7,8]. Still, the H/C ratio, that is, the energy content, rises significantly during the processing.

Water is not likely to react with organic compound at ambient conditions. However, the properties of water change as the temperature increases. The oxygen atom becomes less negative, and the molecule less polar. This is reflected in the dielectric constant, ε, that is reduced from 78.9 to 19.7 when the temperature increases from 25 to 300°C [7,8]. The affinity to essentially non-polar organic molecules increases accordingly. A second critical parameter is the dissociation constant of water, K_w, to H^+ and OH^-. The increase in the equilibrium constant with temperature is shown in Fig. 6.5 as pK_w ($-logK_w$) [7,8]. Changing the temperature from 25 to 250°C increases the dissociation constant by three decades, or a factor of 1000 [7,8]. The rate of both acid and basic reactions is enhanced far beyond the temperature effect on the reaction rate constants. Add to this a limited increase in K_w with pressure, and a substantial increase in hydrolysis by H^+ of organic molecules results.

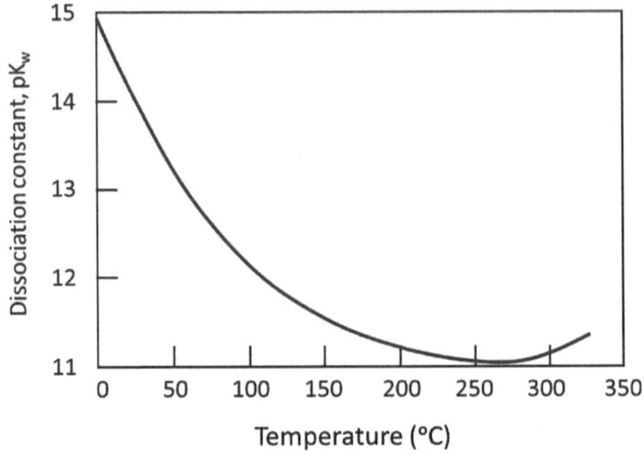

FIGURE 6.5 The dissociation constant of water as a function of temperature at 1 bar. *Data obtained from Ref.* [9].

HTL PROCESSING

Fig. 6.6 shows the process from cultivation of algae to biofuel. Producing algae from municipal wastewater addresses some concerns as to supply of water and nutrients, land use, lifecycle GHG emissions, and handling of waste [10]. Aqueous coproducts, essentially oxygenates that have not been converted, are recycled to the cultivation step. The only products are biofuel and solid biochar. HTL gives a yield of biofuel that

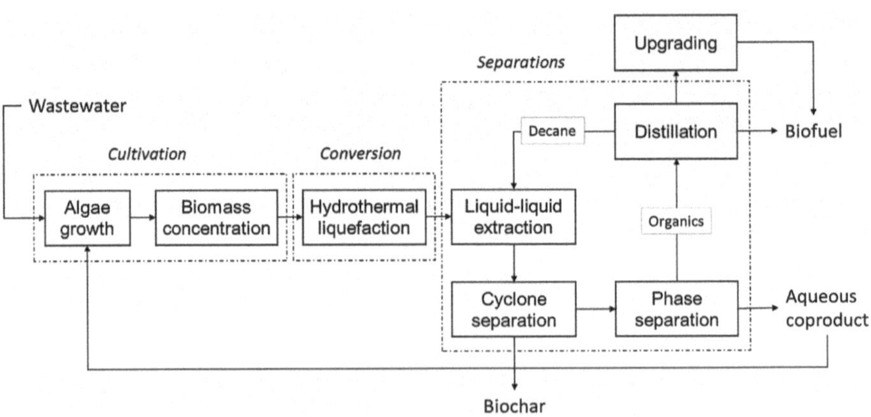

FIGURE 6.6 Biorefinery for biofuel from algae. *Based on Ref.* [11].

is 5%–30% higher than lipid oil extraction that traditionally has been investigated [7,8,10].

In a demonstration in open ponds by the University of Kentucky, algae were produced that contained 29% solid ash (dry weight; dw), 49% carbon (ash-free dry weight; afdw), 38% afdw oxygen, and 14% afdw lipid [10]. The hydrothermal liquefaction was conducted at 350°C and pressures up to 130 bar for 1 hr, giving a yield of biofuel of 44.5% afdw [10]. Weight fraction composition in the produced oil was 78.7% carbon, 10.1% hydrogen, 4.4% nitrogen, and 5.5% oxygen with an energy content of 39 MJ/kg [10]. The analysis showed the fuel to contain a significant number of straight-chain and branched hydrocarbons and mono- and polyaromatics in addition to fatty acids [10]. The solid biochar contained >20% dw carbon with an energy density between 8 and 10 MJ kg^{-1} [10].

The above example only illustrates what typically can be achieved by HTL and the process steps involved. Note that quite an extensive separation and upgrading are involved in order to produce a fuel that is compatible with present specifications. For algae as feedstock expressly, widespread land use has also been prohibited to now for commercialization. HTL can be used to convert kerogen to oil. Kerogen is biological material found in geological formations that yet has not been converted to crude oil. There is considerably more kerogen in the world than crude oil.

A Canadian company has brought the HTL technology further and into the supercritical region using a homogeneous catalyst[b] [12]. They claim 45% recovery of biomass to biocrude, must be dry weight basis, keeping 85% of the energy content in the feed. Overall energy efficiency is ≈80% with the same reduction in GHG emissions compared to fossil fuel. The produced oil has an energy content of 38+ MJ/kg and is blended with crude oil before upgrading in an oil refinery; alternatively, in dedicated on-site processes.

Direct hydroprocessing

Pyrolysis, described in Chapter 3, is a heating process of dried biomass to produce pyrolysis gas and pyrolysis oil. It is different from hydro-

[b] Norwegian Statkraft and Södra Cell of Sweden have announced a demonstration facility in Tofte, Norway, based on the Canadian Steeper Energy technology to be in operation from 2019. Wood residues are going to be the feedstock.

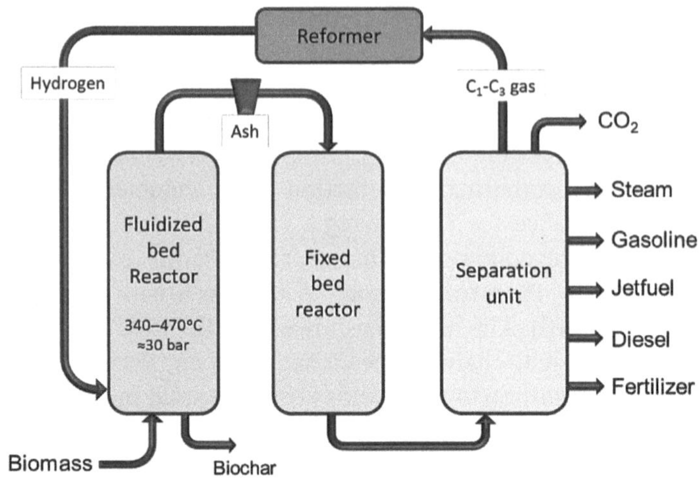

FIGURE 6.7 The IH² hydroprocessing concept. *Based on information from Ref.* [13].

thermal liquefaction in the dry environment and higher temperature. By adding H_2 gas and a catalyst to the process, the products become more abundant in H_2 with properties more resembling conventional fuels. Double bonds become saturated, and oxygen and sulfur are removed to very low levels.

At least one catalytic thermochemical process has been developed and tested that comes under this category, specifically the IH² process[c] [13]. The basics of this process are shown in Fig. 6.7, notably containing twin-reactors; one fluidized-bed and one fixed-bed. Any biomass can be used, but it needs to be dried to 10%–30% moisture and sized for proper reactivity and ease of feeding. It has not been reported how residual ash and produced solid biochar are separated without blocking the reactors, although this part of the technology is reported to have been verified in a pilot plant. We expect fly-ash to be removed by a cyclone on the exit gas from the first rector, and possibly biochar at the bottom of the reactor. Both reactors operate under normal

[c] IH² was developed by the Gas Technology Institute (GTI) in Illinois, US, and has been licensed to CRI Catalyst Company (CRI) for worldwide deployment. CRI is a subsidiary of Shell Global Solutions. A demonstration unit is in the initial testing stage in India.

hydrotreating conditions with H_2 produced as part of the process from evolved light gases. The fluidized-bed reactor is designed to hydrodeoxygenate the feed (i.e., to remove oxygen and cap-free radicals to produce a stable product for further processing). The fixed-bed reactor is a commercial-type hydrotreating unit with a different catalyst from stage one. The reformer is by the technology provider referred to as an H_2 manufacturing unit.

The gasoline and diesel fractions are water white and fulfill ASTM standards. Cetane improvement has been demonstrated for the diesel making it a valuable blending component. An economic evaluation (2012) of a 2000 tons/day woody biomass plant by a third party indicates a break-even fuel price of 0.53 USD/L with a potential for further reduction when the plant is integrated with an existing site such as paper mill or refinery.

Biomass to liquids

Biomass-to-liquids is a subprocess of the more generic XtL processes, where X is any carbon source. Coal-to-liquids (CtL) and gas-to-liquids (GtL) have been developed industrially since 1925 and can be considered mature technology, though there are still developments going on. Several natural gas-to-liquid plants have been built in Qatar, Malaysia, and Nigeria with Shell and Sasol technologies during the last three decades. However, despite several demonstration plants, there is no BtL plant operating today. Fundamental of the XtL technologies is Fischer-Tropsch synthesis (FTS) where hydrocarbon chains in the fuel are being built from monomeric carbon units [14]. The principle is shown in Fig. 6.8.

First, the hydrocarbons or carbohydrates are broken down to small molecular species as syngas in a reformer or gasifier. Heat is added externally, or internally in the reactor by partial combustion, to drive the reaction. Temperature is from 1000–1600°C depending on technology and feedstock. The pressure is between 20 and 40 bar, but atmospheric gasification of biomass has also been demonstrated. Fischer-Tropsch (FT) technology is split into high-temperature (about 350°C) and low-temperature synthesis (about 220°C) [15]. Low-temperature synthesis with a cobalt-based catalyst is most relevant for jet fuel and diesel production due to high selectivity to paraffin [16]. Different reactor

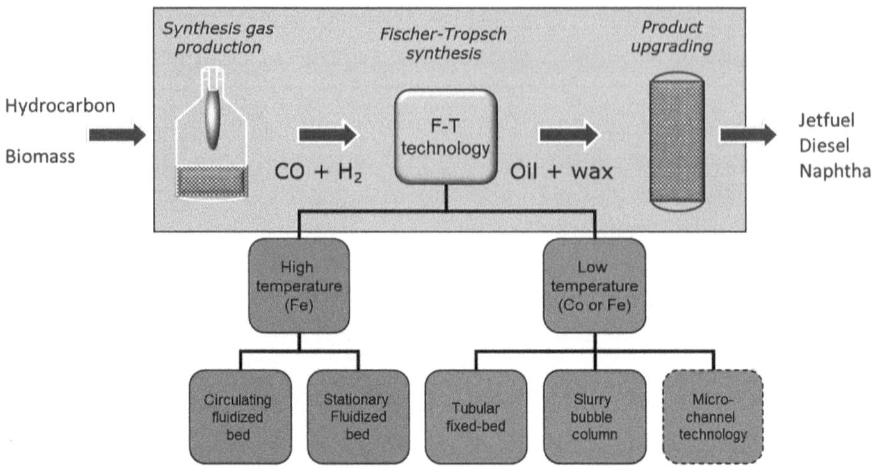

FIGURE 6.8 The XtL principle of fuel production.

technologies are discussed further. After making long-chain hydrocarbons, these are upgraded to fuel by hydrotreating, hydrocracking, and hydro isomerization much in the same way as for HVO.

FISCHER-TROPSCH SYNTHESIS

Fischer-Tropsch synthesis is a polymerization reaction from syngas. Carbon atoms of CO are first hydrogenated to give CH_x monomer species that couple together on the surface of a catalytic metal. Both cobalt and iron are being used; ruthenium is also an FT metal, but too expensive. The products with a cobalt-based catalyst are mainly straight-chain paraffins and α-olefins:

$$n\text{CO} + (2n+1)\text{H}_2 \rightarrow \text{C}_n\text{H}_{2n+2} + n\text{H}_2\text{O} \quad (6.1)$$

$$n\text{CO} + 2n\text{H}_2 \rightarrow \text{C}_n\text{H}_{2n} + n\text{H}_2\text{O} \quad (6.2)$$

A wide range of chain lengths is produced as determined by the value of chain termination probability relative to chain growth probability. In general, the product slate follows the Anderson-Schultz-Flory (SFA) distribution as expressed by:

$$\frac{W_n}{n} = (1-\alpha)^2 \alpha^{n-1} \quad (6.3)$$

where W_n is the weight fraction of a chain with a given chain length and α defines the chain growth probability according to:

$$\alpha = \frac{r_p}{r_p + r_t} \tag{6.4}$$

Here, r_p and r_t are the reaction rates for propagation and termination, respectively. To minimize the production of light gases, it generally will be desirable to have as high α as possible, defined by the actual catalyst used and the process conditions. The number average molecular weight varies as:

$$M_W(\alpha) = \frac{14.026}{(1-\alpha) + 2.016} \tag{6.5}$$

resulting in molecular weight averages for α = 0.90; 0.95 and 0.999, respectively, of 100; 285 and 14,000.

The H_2/CO usage ratio is in the range 2.1 to 2.15 depending on α and to some extent on the selectivity to other products but alkanes. This ratio is higher than generally obtained by gasification of biomass, and the ratio must, therefore, be adjusted before synthesis.

Further details of the mechanism, but still simplified, is shown in Fig. 6.9. Only hydrocarbon formation is shown, as only minor

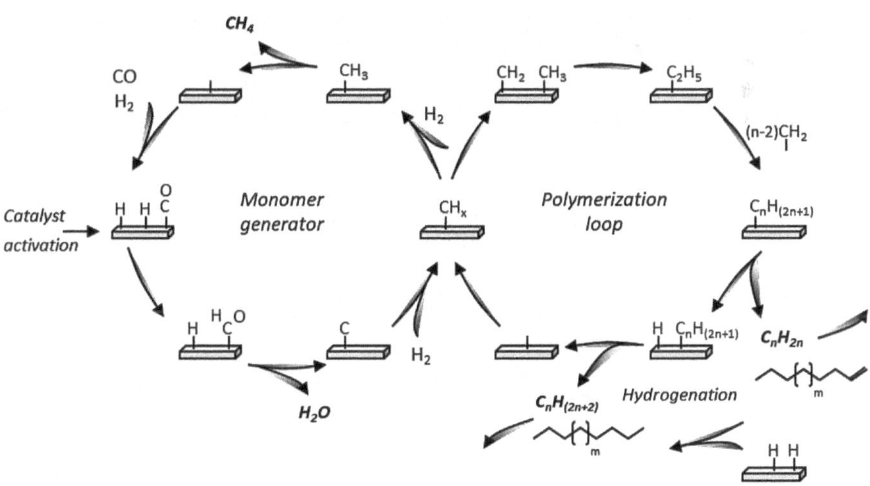

FIGURE 6.9 Fischer-Tropsch mechanism.

amounts of oxygenates typically will be formed. Most of the CH_x monomers generated on the surface of the catalyst are often assumed to be in the form of CH_2 that is readily incorporated in the chain during polymerization. A smaller portion of the monomer is hydrogenated to methane. The growing chain can be terminated by β–hydrogen abstraction and leave the surface as an α-olefin or being hydrogenated to an alkane before termination. Olefins can also be hydrogenated in a secondary reaction. There is evidence from experiments at low conversion and small catalyst particle size that the primary product is dominated by olefins, but at standard process conditions, the olefin-to-paraffin ratio is above two for C_3 and then diminishing rapidly with chain length.

A large number of kinetic expressions for the reaction has been developed. An important finding in recent years is that water produced during the reaction plays a significant role. It has been proposed that water is involved in the dissociation of CO to create CHx. Up to a particular concentration, water improves selectivity to long-chain paraffins and, for most catalysts, improves the reaction rate. For huge water concentrations (i.e., high conversions of CO), the activated metallic cobalt catalyst can re-oxidize, thereby deactivating the catalyst. Example of a cobalt catalyst is shown in Fig. 6.10.

Employed reactor technologies for FTS by Shell and Sasol are tubular fixed-bed and slurry bubble-column, respectively (Fig. 6.8). Other reactor technologies have been tested extensively on a smaller scale, notably the fixed-bed microchannel reactor. These reactors have several pros and cons, encompassing the well-proven tubular fixed-bed

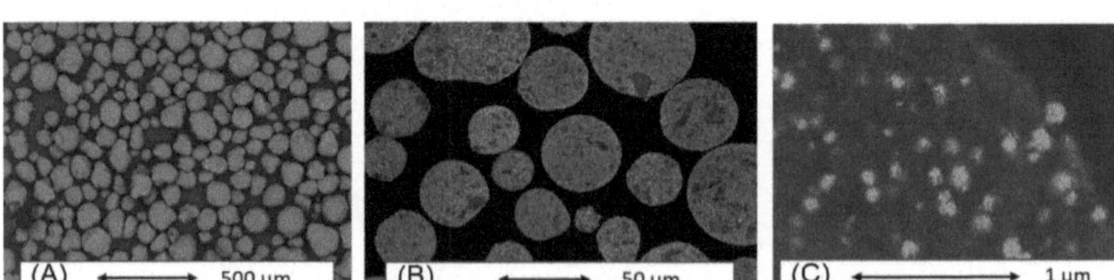

FIGURE 6.10 The catalyst with cobalt supported on γ-alumina for operation in a slurry reactor. (A) γ-alumina particles. (B) Cross section of catalyst particles. (C) Distribution of cobalt agglomerates (*green*) in γ-alumina pores.

with high scalability and excellent heat removal capacity. The slurry bubble-column reactor is characterized by close to full back-mixing of the gases and liquid products, meaning that the catalyst is exposed to the exit gas composition throughout the entire reactor. This has the advantage of improved selectivity to higher hydrocarbons due to a low H_2/CO ratio and high water vapor pressure but comes at the expense of lower partial pressure of the reactants. The CO conversion in an FT-slurry reactor is limited to the range 60%–65%. Important limiting parameters in the design are heat-removal, linear gas velocity, catalyst activity, and reactor length.

THE BtL PROCESS

A block-flow diagram of a standard BtL process is shown in Fig. 6.11. The biomass is first dried and pretreated by pyrolysis or torrefaction. Gasification takes place at an elevated pressure of 10–40 bar, which is advantageous for equipment size (costs), thermodynamics, and kinetics [14]. Gas from pretreatment is injected into the gasification reactor through fuel nozzles and mixed with oxygen and possibly steam. Char is fed through a lock-hopper and a fluidization vessel. Note that all the heat needed for pyrolysis or torrefaction comes from the FT-reactor tail gas, whereas drying is supported by steam from raw syngas cooling within the clean-up block. After gasification, the

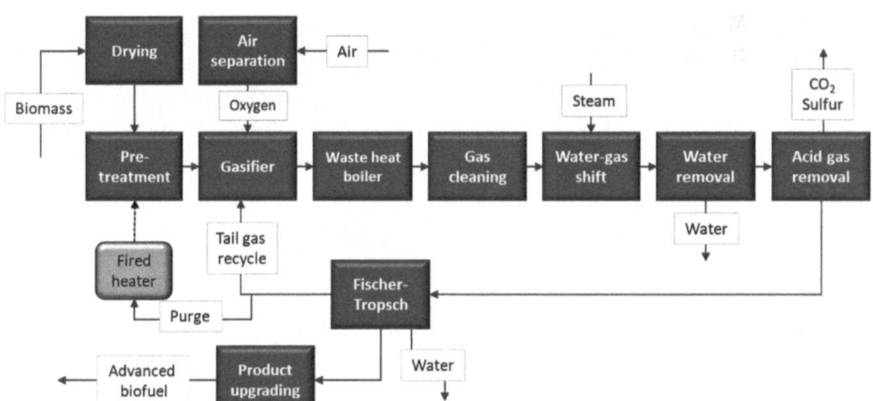

FIGURE 6.11 Block-flow diagram of a conventional BtL process.

produced syngas is quenched with water, thereby producing high-pressure steam. Quenching is necessary to protect construction materials from corrosion, called metal dusting, through the critical temperature range down to ca. 600°C [14]. Cyclone and filters clean the gas before the syngas composition is adjusted to the desired H_2/CO ratio in a catalytic fixed-bed reactor through the water-gas-shift (WGS) reaction (400–220°C):

$$CO + H_2O \rightarrow CO_2 + H_2 \tag{6.6}$$

The WGS reaction is often divided into high-temperature WGS at ca. 400°C using an iron-based catalyst and low-temperature WGS with a copper-based catalyst at 220°C. Also, additional CO_2 is produced, thereby decreasing carbon efficiency of the BtL process [14]. After knocking-out water, the syngas is further cleaned by hydrolysis to convert HCN to NH_3 and COS to H_2S, as well as the removal of CO_2, NH_3, and H_2S. Some CO_2 will be recycled to the gasifier [14]. H_2S is further converted to elemental sulfur for deposition. NO_x and H_2S have to be removed to a concentration of less than 10 ppb.

After all this effort to produce very clean syngas with the right stoichiometric composition, the gas is ready for FT synthesis. The hydrocarbons produced vary in chain length, but usually the FT process is optimized to give a high-wax fraction. This wax is upgraded to the desired products as described further. For the upgrading, particularly the removal of oxygenates and saturation of olefins, a small PSA (pressure swing adsorption) unit is installed at an appropriate location to extract a minor amount of H_2 from syngas. The inlet H_2/CO ratio to the FT reactor(s) preferentially is lower than the overall consumption ratio of ca. 2.15 and is typically 1.7–1.9, meaning that this ratio becomes even lower as the reaction proceeds. The rationale for this choice of conditions is that the selectivity to higher hydrocarbons strongly depends on the H_2/CO ratio, and it is desirable to suppress the formation of methane and light hydrocarbons.

The FT tail-gas directed to the fired-heater also serves as a purge for recycle streams in order not to accumulate inert gases. Recycle around the FT reactor is needed to secure 90% or above conversion of the syngas. Additional tail-gas can be directed to the gasification unit.

BtL PRODUCTS

The BtL diesel is very similar to HVO. Upgrading of FT paraffins (molten wax) is slightly more complicated due to more substantial variation in the chain length. Fig. 6.12 shows a general layout of the upgrading of FT hydrocarbons for maximum diesel production for a plant capacity of 2000 barrels per day (bpd) of liquid products. We notice that there are significant amounts of tail-gas, which is partly used as fuel or regasified, but most are produced water that is knocked out and unconverted syngas that is recycled to the FT reactor. FT hydrocarbons, mostly wax, are first hydrotreated to remove oxygenates and saturate olefins, then the longer chains are cracked and isomerized to give the appropriate product slate and quality. The hydro-treatment can be conducted by using two reactors, or beds, with specific catalysts. The process can be optimized for other products like base oil or the intermediate boiling point range kerosene.

One part of the BtL products, that is, of particular interest, is the kerosene fraction that can be optimized to fulfill the Jet A/A-1 specifications ASTM D1655/D7566 for drop-in aviation fuel. Apart from hydrotreated vegetable oil (HVO) from oil seed plants, the biomass-derived FT product is the only fuel able to reduce greenhouse gas emissions of aircraft on a short to medium term. Blends of up to 50% with FT-BtL jet-fuel have been certified. Several criteria as to thermal stability, cold-properties, viscosity, and combustion properties have to be fulfilled. More than 10 flight tests on large commercial or military planes and blends of up to

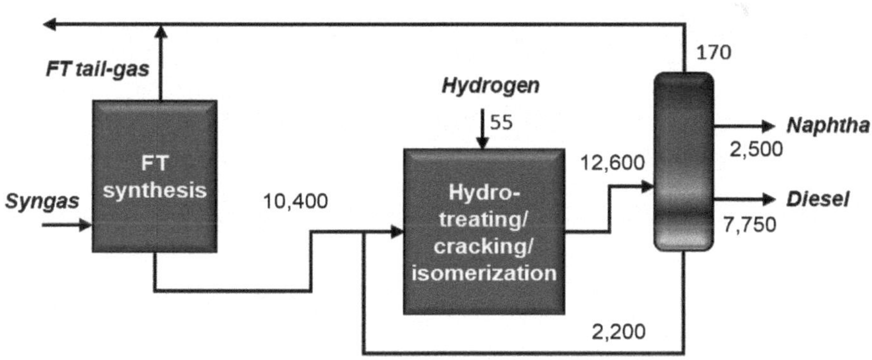

FIGURE 6.12 Upgrading section of a 2,000 bpd BtL plant. Flows are in kg/h.

50% FT jet-fuel have been performed, but all are based on natural gas (GtL) or coal (CtL) as feed. The first commercial flight was conducted by Qatar Airways on an Airbus A340–600 flight from London Gatwick to Doha on October 12, 2009. HVO fuel was certified as late as in June 2011 but has received much attention due to its availability, and several test flights have already been completed. However, for large-scale utilization, HVO has challenges with possible competition for land use and a less efficient GHG reduction impact.

IMPROVING THE EFFICIENCY WITH ADDED HYDROGEN (PBtL)

The conventional BtL process is rather inefficient in terms of energy efficiency and utilization of the carbon in the biomass. A comparison depending on technology selection, shown in Table 6.4, is based on process simulations reported by Statoil. The table shows that only 1/3 of the carbon is converted to usable products. If carbon efficiency could be increased to above 90%, it means that only 30%–40% of the original biomass volume needs to be collected per unit fuel produced.

High carbon utilization is based on converting CO_2 with H_2 to reactive CO; that is, the reverse of the WGS reaction. Details of this reverse reaction are shown further. The main message is that a high temperature is needed for a decent conversion; at least 900°C. Several process units operating at high temperatures open up for integration with a high-temperature SOEC for steam electrolysis. SOEC is described in Chapter 3, and the principle of adding H_2 to BtL is shown in Fig. 6.13.

TABLE 6.4 Efficiencies of Fuel Production in the BtL Process

Pretreatment	Gasifier	Carbon Efficiency (%)	Energy Efficiency (%)
Milling	Circulating fluidized bed	24	27
Torrefaction	Entrained flow	33	27
Pyrolysis	Entrained flow	38	33

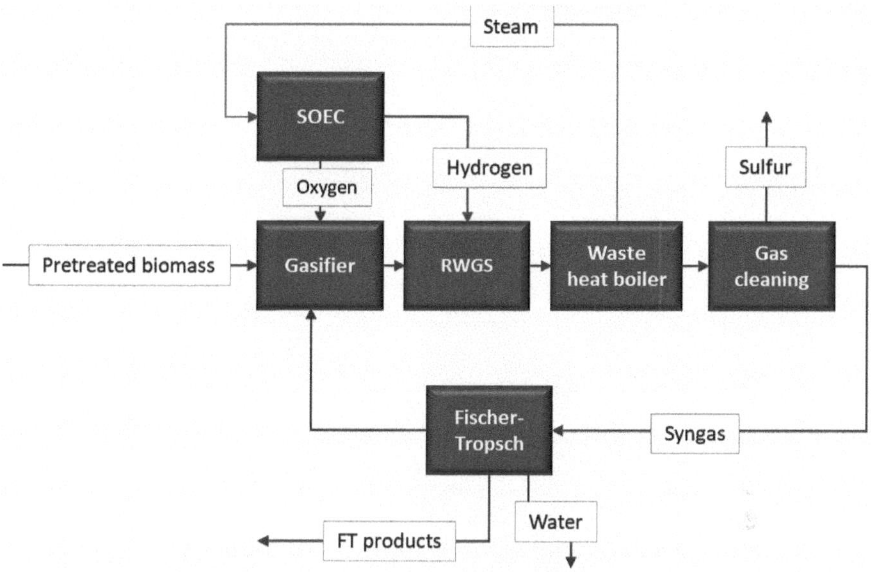

FIGURE 6.13 Block-flow sheet for PBtL high-efficiency fuel production from biomass through steam electrolysis.

The concept is sometimes called PBtL as it, in effect, is adding power (P) to BtL. Note the added benefit of coproduction of oxygen in the SOEC that is used for gasification. A costly air separation unit has been removed. Steam for electrolysis is produced by cooling syngas in a waste-heat-boiler (WHB); no direct water quenching is needed. The concept is conceivably simple.

Technology readiness and cost of SOEC are the main uncertainties. Although SOEC units are sold commercially, the units are small, and they operate at atmospheric pressure. Further, the produced oxygen is not utilized, but diluted with a carrier gas to avoid having neat, highly reactive oxygen at typically 800°C. There is also a question of a lifetime of the oxide membrane as it sinters with time and has to be exchanged after a few years.

The effect of H_2 is illustrated in Fig. 6.14. Simulations show an almost direct benefit to carbon efficiency and fuel production by adding H_2. This means that if it is found profitable to produce H_2, then the maximum amount needed should be produced. There is a practical

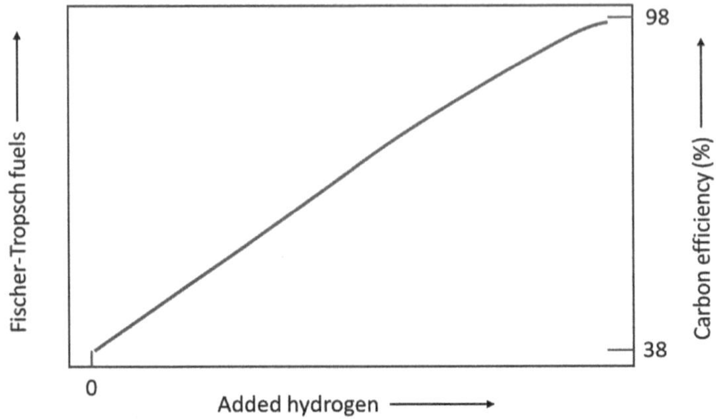

FIGURE 6.14 Effect of adding H_2 to the BtL process.

maximum to carbon efficiency of 97%–98%, as some CO_2 may follow the impurities during gas cleaning, and there is always the need for purge gas to remove inert gases. This purge gas is energy-rich and is used in the fired heater(s) to heat gases where needed (e.g., the feeds to the gasifier).

Hydrogenation of bio-CO_2 (PBCtL)

Hydrogenation of CO_2 that has its origin in biomass is probably the most spectacular example of fuels produced from biomass. Significant energy input is needed in the form of H_2 because CO_2 is at the lowest point of Gibbs free energy of carbonaceous compounds. A comparison of ΔG_f compared with some other C_1 compounds is shown in Fig. 6.15. Only the aqueous carbonate ion is more favorable, covering inorganic carbonates and the organic compound where CO_2 is incorporated like organic acids, polycarbonates, and polyurethanes. Making hydrocarbons, specifically fuels, and simultaneously avoiding GHG emissions from fossil resources, requires the input of solar, wind, geothermal, hydroelectric, or tidal wave energy.

The CO_2 can come from landfill, fermentation, or gasification of any biomass. As long as H_2 is produced using renewable energy, the GHG impact is negative. However, first, CO_2 must be made reactive.

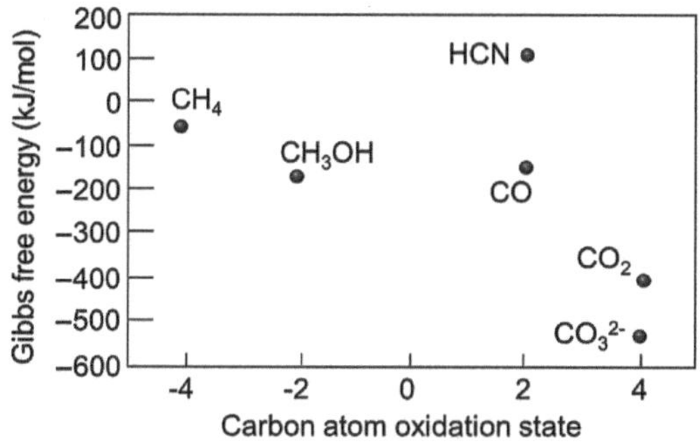

FIGURE 6.15 Gibbs free energy of formation of C_1 species.

MAKING SYNGAS FROM CO_2

The critical step is making CO from CO_2 through the reverse of the water-gas-shift (RWGS) reaction ($\Delta H = 37$ kJ/mol at 1000°C):

$$CO_2 + H_2 \rightarrow CO + H_2O \tag{6.7}$$

How much CO_2 is converted to CO is dictated by the equilibrium in addition to the kinetics of the reaction. As the reaction is endothermic, the reaction is shifted toward CO at increasing temperature and at least 800°C is necessary to have a useful conversion. The equilibrium constant is one at 1100 K. Homogeneous gas phase reactions are reasonably fast at these temperatures, but a catalyst to speed up the reaction rate might be needed. Nevertheless, use of a catalyst is to be explored further as no commercial unit yet has been erected. An example of both equilibrium and kinetics is found in Fig. 6.16 starting at 1000°C with molar gas composition $CO_2 + 3H_2$. This is the approximate overall gas composition needed for both Fischer-Tropsch and methanol synthesis. An adiabatic reactor is assumed (i.e., the reactor is thermally insulated with no heat exchange with the surroundings). The actual composition will be slightly different due to the total process configuration, including recycle of unconverted gas. We notice that it takes at least 20 s before equilibrium is approached and that the temperature during this period drops by 50°C to account for the positive enthalpy of reaction. About 43% of CO_2 in the feed is converted to CO, pointing at the need for unconverted CO_2 to be recycled to the RWGS reactor.

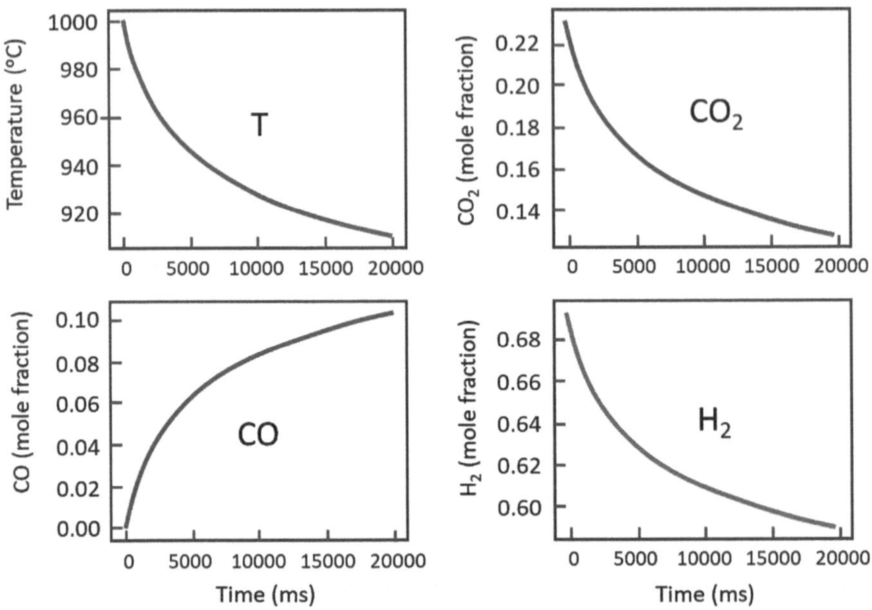

FIGURE 6.16 Kinetics and equilibrium of the RWGS gas phase reaction at 1000°C.

THE PROCESS OF MAKING FUEL

The process is similar to BtL, but it is an absolute requirement to produce and add H_2 when starting with CO_2. The process flowsheet is a further development and simplification of the one shown in Fig. 6.13. No purification is needed after the RWGS as it is assumed that CO_2 is delivered to site according to specifications. This is a challenge for biomass-based CO_2, especially for biogas from the landfill of municipal waste. Apart from the core FT synthesis, there are two main process units: electrolysis of water and RWGS. A lot of energy is needed to heat CO_2 and H_2 for RWGS to be run at ca. 1000°C. The choice of electrolysis technology is not apparent and depends heavily on options for an effective heat exchange network. Maturity of technology and integration with nearby processing plants are other important parameters.

A suggested block-flow diagram is shown in Fig. 6.17. After the Fischer-Tropsch synthesis, there are liquids products, but also a significant

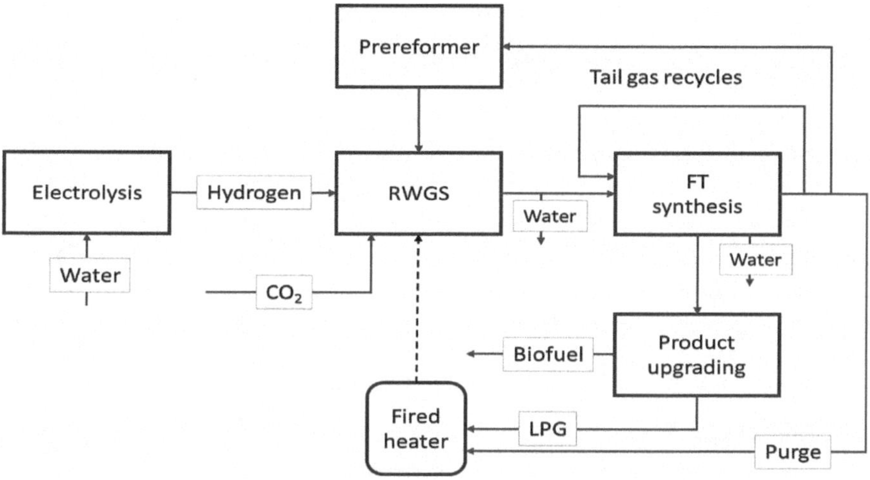

FIGURE 6.17 Developed PBCtL block-flow diagram for converting CO_2 with H_2 to fuel.

gas fraction. This gas consists of unconverted CO and H_2, CO_2 that was not shifted to CO, and produced methane and light hydrocarbons. These components are reduced or removed by several means. First, part of the tail gas is recycled around the FT reactor to improve the conversion of syngas. A second part of the tail gas is directed to the RWGS to convert CO_2. However, this gas contains some light hydrocarbons that are prone to coking at the inlet of the shift reactor. Therefore, a prereformer has been added. This reformer is run at ca. 400°C with a nickel-based catalyst. A typical reaction is:

$$C_3H_8 + CO_2 \rightarrow 2CH_4 + 2CO \tag{6.8}$$

There is also a significant amount of methane (around 10%–15% on a carbon basis) that is made partly during the FT synthesis and partly by the reaction above. Methane is removed through the purge stream that, together with light LPG (Liquid Petroleum Gases) components from hydrocracking in the upgrading section, is directed to a fired heater. This energy is used to heat CO_2 and H_2 to the operating temperature of the RWGS reactor. We notice that the process appears well-integrated, but detailed analyses are needed.

Merging H_2 production with CO_2 reduction and renewable energy eliminates issues with storing, transportation, and use of H_2 as an energy carrier. Conventional combustion engines and present distribution systems for fuels can be used. There are no concerns with mileage or filling. A further option is to use H_2 conversion with CO_2 as temporal storage of electric energy when the demand is low during the night. Windmills can run whenever the wind is blowing, also when the grid is saturated, and the electricity used for electrolyzing water and producing hydrocarbons. Storage is highly efficient as the energy density of hydrocarbons is ca. 35 GJ/m^3; 8 times that of the best lithium-ion batteries on a volume basis and 25 times on a weight basis.

REFERENCES

[1] Agrawal R, Singh NR, Ribeiro FH, Delgass WN. Sustainable fuel for the transportation sector. Proc Natl Acad Sci 2007;104(12):4828–33.
[2] Hannula I. Hydrogen enhancement potential of synthetic biofuels manufacture in the European context: a techno-economic assessment. Energy 2016;104:199–212.
[3] Neste. NEXBTL technology is our own innovation. Available from: https://www.neste.com/corporate-info/who-we-are/research/nexbtl-technology [accessed: 12.04.19].
[4] Weber C, Amundsen AH. Renewable fuels - renewable diesel fuel: HVO. Available from: https://www.toi.no/getfile.php/1343045/Publikasjoner/T%C3%98I%20rapporter/2016/1475-2016/1475-2016-elektronisk.pdf; 2016.
[5] Neste (2017). Renewable diesel handbook [Online]. Available from: https://www.neste.com/sites/default/files/attachments/neste_renewable_diesel_handbook.pdf.
[6] Elliott DC. Historical developments in hydroprocessing bio-oils. Energy Fuels 2007;21(3):1792–815.
[7] Gollakota ARK, Kishore N, Gu S. A review on hydrothermal liquefaction of biomass. Renew Sustain Energy Rev 2018;81:1378–92.
[8] Elliott DC, Biller P, Ross AB, Schmidt AJ, Jones SB. Hydrothermal liquefaction of biomass: developments from batch to continuous process. Bioresour Technol 2015;178:147–56.
[9] IAPWS. International Association for the Properties of Water and Steam. Available from: http://www.iapws.org/; 2019 [accessed 12.04.19].
[10] Roberts GW, Fortier M-OP, Sturm BSM, Stagg-Williams SM. Promising pathway for algal biofuels through wastewater cultivation and hydrothermal conversion. Energy Fuels 2013;27(2):857–67.
[11] Suganya T, Varman M, Masjuki HH, Renganathan S. Macroalgae and microalgae as a potential source for commercial applications along with biofuels production: a biorefinery approach. Renew Sust Energ Rev 2016;55:909–41. doi:10.1016/j.rser.2015.11.026.

[12] Steeper Energy. Hydrofaction™. Available from: https://steeperenergy.com/hydrofaction/; 2018 [accessed 12.04.19].
[13] CRI Catalyst Company & Shell. CRI IH2. Available from: https://s04.static-shell.com/content/dam/shell-new/local/business/cri-catalyst/downloads/pdf/cri-ih2.pdf; 2018 [accessed 12.04.19].
[14] Rytter E, Ochoa-Fernández E, Fahmi A. Biomass-to-liquids by the Fischer–Tropsch process. Catal Process Dev Renew Mater 2013;10:2.
[15] Schulz H. Short history and present trends of Fischer–Tropsch synthesis. Appl Catal A Gen 1999;186(1–2):3–12.
[16] Rytter E, Holmen A. Perspectives on the effect of water in cobalt Fischer–Tropsch synthesis. ACS Catal 2017;7(8):5321–8.
[17] Lamb JJ, Islam MH, Hjelme DR, Pollet BG, Lien KM. Effect of power ultrasound and Fenton reagents on the biomethane potential from steam-exploded birchwood. Ultrason Sonochem. **58**, 2019;104675. Available from: http://www.sciencedirect.com/science/article/pii/S1350417719308909.

Chapter | seven

Promising Selected Biohydrogen Solutions

Jacob J. Lamb[a], Kristian M. Lien[b]

[a]*Department of Electronic Systems, Department of Energy and Process Engineering, ENERSENSE, Faculty of Information Technology and Electrical Engineering, NTNU, Trondheim, Norway;* [b]*Department of Energy and Process Engineering, ENERSENSE, Faculty of Engineering, NTNU, Trondheim, Norway*

CHAPTER OUTLINE

ABSTRACT

Biological hydrogen production is a well-studied area. Despite this, the low yields achieved have prevented such technologies from becoming widespread for hydrogen production. Moreover, low yields also lead to the generation of side products whose large-scale production would generate a waste disposal problem. Of the possible approaches that are under active investigation (biophotolysis, photofermentation, microbial electrolysis, and dark fermentation), dark fermentation has the additional advantage of largely relying on already developed bioprocess technology and of potentially using various waste streams as feedstock. Biological hydrogen production presents a possible avenue for the large-scale sustainable generation of hydrogen needed to fuel a future hydrogen economy. In this chapter, recent attempts to overcome these barriers are briefly discussed and recent progress in efforts to increase biological hydrogen yields are described.

Hydrogen, Biomass and Bioenergy. http://dx.doi.org/10.1016/B978-0-08-102629-8.00007-4

CURRENT BIOHYDROGEN STATUS

There has been a large amount of effort made to understand further and improve biohydrogen (bio-H_2) production. Significant advancements and breakthroughs have been reported concerning microorganisms and their adaptation to environmental changes [1], as well as the metabolic mechanisms involved, and how they are influenced by external factors [2]. Despite these efforts, many technical and physiological barriers must be addressed to allow the industrial implementation of these bio-H_2 approaches (Table 7.1).

Despite many microorganisms being able to produce H_2, the yields are low as there is competition with other metabolic pathways within the organism for ferredoxin and NAD(P)H [3]. The metabolic pathways involved with H_2 evolution in microorganisms are also very complicated, due to the high diversity of hydrogenases within individual organisms performing different metabolic functions [4]. Genetic engineering has been used as a potential tool to improve H_2 production from microorganisms in recent years. To achieve this, alternative metabolic pathways that compete with H_2 production have been genetically blocked [5]; however, in many of these attempts, observations of improvements in H_2 production have not been achieved. Although genetic modification could eventually improve the H_2 production in microorganisms, the use of genetically engineered microorganisms in large-scale plants would result in a significant increase in operational costs and increase environmental concerns regarding the use of genetically altered organisms [6]. Additionally, equipment for live maintenance of the optimum culture density is also required for optimum production [7,8].

Fermentive biohydrogen production

Despite scientific and technological improvements, an economically beneficial solution to bio-H_2 production is yet to be achieved. Fermentative bio-H_2 production is an example of this, where the research has focused on a limited number of microorganisms in dark fermentation systems with a synthetic, glucose-rich substrate [9,10]. The observations made can help elucidate metabolic dynamics in the specific organisms, but has limited application in an industrial process, due to the complexity of the substrates used that introduce

TABLE 7.1 Various Biohydrogen Production Processes [12]

Process	Strengths	Weaknesses
Direct biophotolysis by eukaryotic green algae	Has direct H_2 production from water and sunlight, with higher solar conversion and lower nutrient requirement than observed in terrestrial plants, and in many cases, they can fix N_2 from the atmosphere	Have a low-light utilization efficiency, hard to promote H_2 production and therefore the low rate of production, and are highly oxygen sensitive
Indirect biophotolysis by cyanobacteria	Has hydrogenase H_2 production using photosynthetically derived carbohydrates, with higher solar conversion and lower nutrient requirement than observed in terrestrial plants	Have a low-light utilization efficiency, hard to promote H_2 production and therefore the low rate of production, and are highly oxygen sensitive
Photo-fermentation by non-sulfur photosynthetic bacteria	These organisms can use a broad spectrum of light, a variety of organic waste effluents, and both hydrogenase and nitrogenase-based H_2 evolution is possible	The fermentation process is easily contaminated by H_2-consuming organisms, light and dark cycle utilization has not been perfected, reactor design for optimal light penetration, and how to integrate these into a biorefinery
Dark fermentation	Can continuously produce H_2 without light, a variety of carbon sources can be utilized as substrates, valuable metabolites (butyric, lactic, and acetic acid) are produced as by-products, and is entirely anaerobic	Genetic systems and reactor design require further understanding and development; biomass convertibility is poorly understood, extraction of H_2, and the presence of CO_2 in the gas mix
Microbial electrolysis cells	Potential for high yields of H_2, biochemical oxygen demand can be reduced due to the ability to use various effluents as reaction substrates and is well-suited for integration with other bioprocesses	The requirement for microorganisms to operate at a lower pH and higher temperature, potential losses at the cathode, slow proton transport, ion exchange membranes are expensive components, and therefore these systems are hard to scale-up

further microorganisms into the system [11]. This can lead to undesired metabolic products other than H_2 (methane and acetate), and increased levels of acids and biochemical oxygen demand, resulting in the requirement for extensive effluent treatment after the fermentation process.

To alleviate the requirement for post-fermentation treatment, research has begun examination of using the fermentation effluent as a source of further organic substrates for separate and combined dark and photo-fermentation, as well as mixed bio-H_2 and methane fermentations [13]. These systems incorporate methanogens, or photosynthetic bacteria, in combination with H_2-producing organisms to maximize biofuel generation [13]. This suggested system is far from perfect, as there are still concerns regarding the inhibitory effect of the effluent on photosynthetic bacteria; however, this has opened up new hybrid possibilities for further development. One such hybrid approach involves the integration of dark fermentation with a microbial electrolysis cell (MEC). The effluent from the dark fermentation would act as a cost-effective substrate for a MEC, ultimately generating significant amounts of H_2 in total (dark fermentation and MEC together), as well as treating the effluent before release into the environment [14,15]. However, this process requires significant improvements to achieve cost-effective H_2 generation on an industrial scale.

Photosynthetic biohydrogen production

The interest in photosynthetically derived H_2 production has been increasing in recent years. The evolution of H_2 has been linked directly to solar irradiation, where the photosynthetic machinery utilizes light energy to split water into O_2 and H_2 [16–19].

There is additionally an indirect light-independent pathway to evolve H_2, with a fermentative degradation of carbohydrates. The main thing in common with these H_2-evolving pathways is the reduction of ferredoxin to provide electrons to hydrogenases. Despite this, many challenges limit bio-H_2 production significantly in photosynthetic organisms. This includes the high-oxygen sensitivity of hydrogenases (which is a significant problem in a predominantly oxygen-evolving organism), competition from other metabolic pathways for reductants like ferredoxin, and inefficiencies in solar irradiation capture and utilization

[20]. Although efforts have been made to improve these processes (e.g., oxygen sequestration [21], photosynthetic antenna [22], and machinery [23] mutations), these observations have been made in costly specific laboratory microenvironments, where upscaling would result in insufficient H_2 production economically.

Microbiological electrolysis cells

Microbial electrolysis cells (MECs) are the combination of bacterial metabolism with electrochemistry resulting in the production of H_2 [24]. Anode respiring bacteria (ARB) are attached at the anode, oxidizing simple organic compounds (ethanol, butyrate, lactate, acetate, and propionate) and transferring the electrons to the conductive solid [25–33]. The electrons are then conducted to the cathode where they react with H_2O and produce H_2 in the cathode compartment. Input of external power is required to boost the negativity of the current supplied so that reduction of H_2O to H_2 can be achieved. The standard potential (E^o) of the electron donor is around –0.28 V, and since the standard potential of H_2 is –0.41 V, theoretically –0.13 V must be supplied externally, nine-fold less than what is required for water electrolysis (standard potential of H_2O is about 0.82 V). Despite this, due to energy losses, the externally supplied power must be higher than –0.13 V (usually between 0.6 and 1.2 V), because of losses occurring at the MEC [34–36].

The yield of H_2 obtainable from MEC system is rather high in the range of 67%–91%, from a diverse range of organic substrates as electron donors [35,37]. This, of course, maximizes the energy output but also reduces the biochemical oxygen demand (BOD) of the effluent. Furthermore, MECs seemingly can completely oxidize organic compounds into CO_2 via the citric acid cycle anaerobically and respire the generated electrons to the anode [37]. The combination of MECs with dark fermentation can increase the yield of H_2, as complex organic substrates can be utilized by dark fermentation, then the simple organic substrates remaining in the effluent can be used by ARBs in a MEC to produce H_2 [38,39].

The requirement for an external power source to achieve H_2 production in MECs is the primary challenge. They are energy losses occurring in several regions within the MEC, which can add up to be rather significant. If these losses become too large, the energy required for the production of H_2 becomes more significant than the net energy output. Therefore, the input power must be maintained below –0.6 V to achieve a net positive energy balance [40]. The rate of H_2 production is also an area requiring improvement to maintain low-capital costs. Since the production rate is proportional to the current, it is dependent on the current density at the anode and the anode's specific surface area. The highest H_2 production rate at current is 0.13 L of H_2 per L of culture per h with an applied voltage of 0.8 V [35], compared with up to 7.9 L of H_2 per L of culture per h from dark fermentation [41].

Methane production is inevitable in MEC, because the growth of methanogens is fast, although methane production can be suppressed if the H_2 production rate is high enough [42]. Methane production is usually a problem as it can contaminate produced H_2. However, methane production is intentional because methane is widely used as chemical energy. The production of methane by MEC has several advantages: the product gas has a higher methane content than that of other technologies, no additional heat is required, and wastewater with low organic substrate content can be used, which cannot be treated by anaerobic digestion. However, the Faraday efficiency of this process is currently low (around 25%) [43]. Nafion® is generally utilized as a separator membrane and to transport protons. However, it can induce a pH gradient on both the anode and cathode, resulting in overpotential [42], and the cost of electrolytes for the large-scale operation of this process is expensive.

Dark fermentation can be coupled to MEC for the production of H_2. Dark fermentation itself does not require light, and it requires moderate operation conditions and can utilize various organic substrates with low-energy demand [44]. However, it has a low H_2 recovery of approximately 23%–25% due to thermodynamic limitations [45,46]. To increase efficiency, MEC is utilized to produce H_2 from effluent from the fermentation [39,47]. Light-assisted MEC H_2 production is also possible and focuses on assisting MEC with energy from a solar cell. Dye-sensitized solar cells (DSSC) are preferentially utilized in research regarding this topic. They can achieve a high recovery efficiency of approximately 80% [48]. The additional costs of a solar cell are a challenge, and thus, studies aim to reduce the use of other expensive components such as platinum in cathodes [49].

PROMISING APPROACHES

Novel hybrid solutions

A significant amount of attention has been given to identifying and developing integrated hybrid systems for bio-H_2 production to not only reduce the shortcomings of the previously described production methods but to take advantage of the characteristics of each method to increase overall bio-H_2 yields. An example of this is the integration of the dark fermentation process with MECs. Additionally, by using wastewater as the substrate, the technology could become more economically feasible, as the production of bio-H_2 would be integrated with wastewater treatment [6]. Despite this, there are still many technological barriers that hinder this process, making the scaling-up of MECs economically unfeasible in the foreseeable future [11]. Alternatively, microalgae-driven photo-fermentative bio-H_2 production processes using the effluent from dark fermentation as the organic substrate have made improvements recently [50]. Chandra and Mohan (2011) showed that certain microalgae strains were capable of the continuous evolution of H_2, without pH or oxygen inhibiting the process [50]. They achieved this by promoting functional relationships between the bacterial-archaeal fermentation organisms and the microalgae strains, where the fermentative organisms were able to maintain a low dissolved oxygen environment allowing the microalgae to generate H_2 continuously [50]. Additionally, considerable amounts of high-value products (antioxidants, fatty acids, polysaccharides, glycerol, glycoproteins, antibiotics, antifreeze proteins, and mycosporine-like amino acids), were also produced in some microalgae strains supplementary to H_2 production [51,52]. Many of these products would add additional economic value to energy, nutrition, pharmaceutical, and agricultural industries; therefore, their successful extraction and application could improve the economic feasibility of this bioH_2 production process.

Using a dark fermentation process integrated with a microalgae-driven photo-degradation of the effluent could also result in the fixation of CO_2 produced during dark fermentation and utilized by the microalgae [20,53]. Large-scale implementation of such a system could then theoretically provide a platform for energy production that also aids the reduction of atmospheric CO_2. To date, cost analysis of the implementation of industrial-scale microalgae photobioreactors have been centered on sulfur-deficient bioH_2 technologies, or microalgae biodiesel production [54,55]. The results of such studies suggested that additional highly optimized technological advancements are necessary to achieve a positive energy balance. Furthermore, if the required nutrients, water, and CO_2 could be acquired at a lower economic cost, this could significantly improve the economic feasibility of such a bioH_2 platform [6]. The proposed hybrid dark fermentation integration with a photoheterotrophic degradation system is proposed to have a significantly lower implementation and operation cost compared to highly specific traditional systems [6]. Moreover, the required reduction in nutrients, water, and CO_2 costs could be achieved by using the dark fermentation of wastewater, and subsequent microalgae degradation of the dark fermentation effluent.

Main technological and scientific barriers

For the successful implementation of a hybrid dark fermentation integrated with photoheterotrophic for bioH_2 production, several scientific and technological hurdles must be addressed. One such hurdle is the establishment of an optimal formula for the microbial and microalgae communities used as the initial inoculum [56,57]. Scientific approaches like metagenomics and metabolomics could lead to an understanding of the complexity of the dynamic multi-organism system [58]. A tendency for a shift from a diverse culture to a more particular specialized community has been observed in such multi-organism systems in response to micro-environmental changes [59]. Furthermore, about photoheterotrophic algal bioH_2 production systems, both natural and artificial algal-bacterial associations and communities enhanced the overall H_2 production [60,61].

For significant yields of H_2 to be realized, a tight symbiosis between all organisms and technologies involved is essential to avoid wasting any potentially utilizable component. For example, it is essential that by-products in the effluent of dark fermentation be available to access and be utilized by the microalgae in the next stage [62]. These effluents are laden with organic acids and other useful nutrients; therefore, microalgae strains that can survive and generate significant levels of bioH_2, oils, and other high-value products is essential for this platform to be economically feasible [60]. Additionally, the consumption of considerably large amounts of acetate, volatile fatty acids, and BOD reduction can also be achieved using this platform, displaying that the photoheterotrophic algal platform is also useful for effluent treatment [61]. It was concluded that the transfer of nutrients between various organisms, and their ability to do so while keeping the O_2 levels low, are beneficial for the fermentative and photolytic H_2 evolution by the microalgae. Recently, Boboescu et al. (2015) experimented with using wastewater from beer brewing for dark fermentation and subsequent photoheterotrophic microalgae degradation of the dark fermentation effluent [63]. The study showed that the approach is not only suitable for bioH_2 production but beer production, algal biomass production, and wastewater treatment [63]. Furthermore, light-dark cycles seemingly affected the yields of bioH_2 and algal biomass and tweaking of these cycles and environmental conditions allowed the direction of the process toward either higher bioH_2 production or higher biomass (lipid accumulation) production [50].

The specific conditions of operation of both the dark fermentation and the photoheterotrophic microalgae degradation are essential for optimal performance to be achieved. Moreover, a good comprehension of the influences of various factors within each process, as well as in the overall process, is necessary for the highest possible bioH_2 yields to be achieved. Advanced statistical analysis methods have been constructed for determining areas that can be optimized within technological processes, and these could be applied to this hybrid bioH_2 platform [64,65].

The design and integration of optimal hybrid bioH_2 production require considerable research and development, and the operational and construction costs must be reduced for this hybrid approach to become feasible from an economic point of view [66]. The integration of bioH_2 production with wastewater treatment, CO_2 recycling, and production of high-value products will significantly help increase the value of this hybrid platform, and, therefore, increase the economic feasibility.

REFERENCES

[1] Rafrafi Y, Trably E, Hamelin J, Latrille E, Meynial-Salles I, Benomar S, et al. Sub-dominant bacteria as keystone species in microbial communities producing bio-hydrogen. Int J Hydrogen Energy. 2013;38(12):4975–85.
[2] Greening C, Cook GM. Integration of hydrogenase expression and hydrogen sensing in bacterial cell physiology. Curr Opin Microbiol. 2014;18:30–8.
[3] Mazzoli R. Development of microorganisms for cellulose-biofuel consolidated bioprocessings: metabolic engineers'tricks. Comput Struct Biotechnol J. 2012;3(4):e201210007.
[4] Tran KT, Maeda T, Wood TK. Metabolic engineering of *Escherichia coli* to enhance hydrogen production from glycerol. Appl Microbiol Biotechnol. 2014;98(10):4757–70.
[5] Cai G, Jin B, Monis P, Saint C. A genetic and metabolic approach to redirection of biochemical pathways of *Clostridium butyricum* for enhancing hydrogen production. Biotechnol Bioeng. 2013;110(1):338–42.
[6] Boboescu IZ, Gherman VD, Lakatos G, Pap B, Bíró T, Maróti G. Surpassing the current limitations of biohydrogen production systems: the case for a novel hybrid approach. Bioresour Technol [Internet]. 2016;204:192–201. Available from: http://www.sciencedirect.com/science/article/pii/S096085241501723X.
[7] Kutschera A, Lamb JJ. Cost-effective live cell density determination of liquid cultured microorganisms. Curr Microbiol. 2018;75(2):231–6.
[8] Lamb JJ, Eaton-Rye JJ, Hohmann-Marriott MF. A cost-effective solution for the reliable determination of cell numbers of microorganisms in liquid culture. Curr Microbiol. 2013;67.(2).
[9] Masset J, Calusinska M, Hamilton C, Hiligsmann S, Joris B, Wilmotte A, et al. Fermentative hydrogen production from glucose and starch using pure strains and artificial co-cultures of Clostridium spp. Biotechnol Biofuels. 2012;5(1):35.
[10] Ren H-Y, Liu B-F, Kong F, Zhao L, Xing D, Ren N-Q. Enhanced energy conversion efficiency from high strength synthetic organic wastewater by sequential dark fermentative hydrogen production and algal lipid accumulation. Bioresour Technol. 2014;157:355–9.
[11] Lee H-S, Vermaas WFJ, Rittmann BE. Biological hydrogen production: prospects and challenges. Trends Biotechnol. 2010;28(5):262–71.
[12] Hallenbeck PC, Abo-Hashesh M, Ghosh D. Strategies for improving biological hydrogen production. Bioresour Technol. 2012;110:1–9. Available from: http://dx.doi.org/10.1016/j.biortech.2012.01.103.
[13] Bičáková O, Straka P. Production of hydrogen from renewable resources and its effectiveness. Int J Hydrogen Energy. 2012;37(16):11563–78.
[14] Logan BE, Call D, Cheng S, Hamelers HVM, Sleutels THJA, Jeremiasse AW, et al. Microbial electrolysis cells for high yield hydrogen gas production from organic matter. Environ Sci Technol. 2008;42(23):8630–40.
[15] Wang H, Ren ZJ. A comprehensive review of microbial electrochemical systems as a platform technology. Biotechnol Adv. 2013;31(8):1796–807.

[16] Lamb JJ, Røkke G, Hohmann-Marriott MF. Chlorophyll fluorescence emission spectroscopy of oxygenic organisms at 77 K. Photosynthetica. 2018;56(1):105–24.
[17] Kutschera A, Lamb JJ. Light meter for measuring photosynthetically active radiation. Am J Plant Sci. 2018;09(12):2420–8.
[18] Lamb J, Forfang K, Hohmann-Marriott M. A practical solution for 77 K fluorescence measurements based on LED excitation and CCD array detector. PLoS One. 2015;10(7):e0132258. doi:10.1371/journal.pone.0132258.
[19] Lamb JJ, Eaton-Rye JJ, Hohmann-Marriott MF. An LED-based fluorometer for chlorophyll quantification in the laboratory and in the field. Photosynth Res. 2012;114(1):59–69. doi:10.1007/s11120-012-9777-y.
[20] Kumar K, Das D.CO_2 sequestration and hydrogen production using cyanobacteria and green algae. In: Razeghifard R, editor. Natural and Artificial Photosynthesis; 2013, doi:10.1002/9781118659892.ch6
[21] Wu S, Xu L, Huang R, Wang Q. Improved biohydrogen production with an expression of codon-optimized hemH and lba genes in the chloroplast of *Chlamydomonas reinhardtii*. Bioresour Technol. 2011;102(3):2610–6.
[22] Kosourov SN, Ghirardi ML, Seibert M. A truncated antenna mutant of *Chlamydomonas reinhardtii* can produce more hydrogen than the parental strain. Int J Hydrogen Energy. 2011;36(3):2044–8.
[23] Scoma A, Krawietz D, Faraloni C, Giannelli L, Happe T, Torzillo G. Sustained H_2 production in a *Chlamydomonas reinhardtii* D1 protein mutant. J Biotechnol. 2012;157(4):613–9.
[24] Liu H, Grot S, Logan BE. Electrochemically assisted microbial production of hydrogen from acetate. Environ Sci Technol. 2005;39(11):4317–20.
[25] Bond DR, Lovley DR. Electricity production by Geobacter sulfurreducens attached to electrodes. Appl Environ Microbiol. 2003;69(3):1548–55.
[26] Gorby YA, Yanina S, McLean JS, Rosso KM, Moyles D, Dohnalkova A, et al. Electrically conductive bacterial nanowires produced by Shewanella oneidensis strain MR-1 and other microorganisms. Proc Natl Acad Sci. 2006;103(30): 11358–63.
[27] Kim HJ, Park HS, Hyun MS, Chang IS, Kim M, Kim BH. A mediator-less microbial fuel cell using a metal reducing bacterium *Shewanella putrefaciens*. Enzyme Microb Technol. 2002;30(2):145–52.
[28] Qiao Y, Li CM, Bao S-J, Lu Z, Hong Y. Direct electrochemistry and electrocatalytic mechanism of evolved Escherichia coli cells in microbial fuel cells. Chem Commun 2008;(11):1290–2. doi:10.1039/b719955d.
[29] Reguera G, McCarthy KD, Mehta T, Nicoll JS. Extracellular electron transfer via microbial nanowires. Nature. 2005;435(7045):1098.
[30] Lamb J.J. The Role of Type IV Pili of Synechocystis sp. PCC 6803 in Extracellular Electron Transfer. University of Otago; 2013.
[31] Lamb JJ, Hill RE, Eaton-Rye JJ, Hohmann-Marriott MF. Functional role of pila in iron acquisition in the cyanobacterium Synechocystis sp. PCC 6803. PLoS One 2014;9(8.):e105761. doi:10.1371/journal.pone.0105761.

[32] Lamb J.J. Characterization of Bacterial Electron Transport to Extracellular Electron Acceptors. Doctoral dissertation, Trondheim: NTNU. Available from: https://ntnuopen.ntnu.no/ntnu-xmlui/handle/11250/2399463; 2016.

[33] Lamb JJ, Hohmann-Marriott MF. Manganese acquisition is facilitated by PilA in the cyanobacterium Synechocystis sp. PCC 6803. PLoS One. 2017;12(10).

[34] Tartakovsky B, Manuel MF, Wang H, Guiot SR. High rate membrane-less microbial electrolysis cell for continuous hydrogen production. Int J Hydrogen Energy. 2009;34(2):672–7.

[35] Call D, Logan BE. Hydrogen production in a single chamber microbial electrolysis cell lacking a membrane. Environ Sci Technol. 2008;42(9):3401–6.

[36] Call DF, Merrill MD, Logan BE. High surface area stainless steel brushes as cathodes in microbial electrolysis cells. Environ Sci Technol. 2009;43(6):2179–83.

[37] Segura D, Mahadevan R, Juárez K, Lovley DR. Computational and experimental analysis of redundancy in the central metabolism of Geobacter sulfurreducens. PLoS Comput Biol. 2008;4(2):e36.

[38] Lalaurette E, Thammannagowda S, Mohagheghi A, Maness P-C, Logan BE. Hydrogen production from cellulose in a two-stage process combining fermentation and electrohydrogenesis. Int J Hydrogen Energy. 2009;34(15):6201–10.

[39] Lu L, Ren N, Xing D, Logan BE. Hydrogen production with effluent from an ethanol–H_2-coproducing fermentation reactor using a single-chamber microbial electrolysis cell. Biosens Bioelectron. 2009;24(10):3055–60.

[40] Lee H-S, Rittmann BE. Characterization of energy losses in an upflow single-chamber microbial electrolysis cell. Int J Hydrogen Energy. 2010;35(3):920–7.

[41] Lo Y-C, Lee K-S, Lin P-J, Chang J-S. Bioreactors configured with distributors and carriers enhance the performance of continuous dark hydrogen fermentation. Bioresour Technol. 2009;100(19):4381–7.

[42] Kadier A, Kalil MS, Abdeshahian P, Chandrasekhar K, Mohamed A, Azman NF, et al. Recent advances and emerging challenges in microbial electrolysis cells (MECs) for microbial production of hydrogen and value-added chemicals. Renew Sustain Energy Rev. 2016;61:501–25.

[43] Zhen G, Kobayashi T, Lu X, Xu K. Understanding methane bioelectrosynthesis from carbon dioxide in a two-chamber microbial electrolysis cells (MECs) containing a carbon biocathode. Bioresour Technol. 2015;186:141–8.

[44] Junghare M, Subudhi S, Lal B. Improvement of hydrogen production under decreased partial pressure by newly isolated alkaline tolerant anaerobe, *Clostridium butyricum* TM-9A: optimization of process parameters. Int J Hydrogen Energy. 2012;37(4): 3160–8.

[45] Ust'ak S, Havrland B, Muñoz JOJ, Fernández EC, Lachman J. Experimental verification of various methods for biological hydrogen production. Int J Hydrogen Energy. 2007;32(12):1736–41.

[46] Manish S, Banerjee R. Comparison of biohydrogen production processes. Int J Hydrogen Energy. 2008;33(1):279–86.

[47] Liu W, Huang S, Zhou A, Zhou G, Ren N, Wang A, et al. Hydrogen generation in microbial electrolysis cell feeding with fermentation liquid of waste activated sludge. Int J Hydrogen Energy. 2012;37(18):13859–64.

[48] Ajayi FF, Kim K-Y, Chae K-J, Choi M-J, Kim S-Y, Chang I-S, et al. Study of hydrogen production in light assisted microbial electrolysis cell operated with dye sensitized solar cell. Int J Hydrogen Energy. 2009;34(23):9297–304.
[49] Chae K-J, Choi M-J, Kim K-Y, Ajayi FF, Chang I-S, Kim IS. A solar-powered microbial electrolysis cell with a platinum catalyst-free cathode to produce hydrogen. Environ Sci Technol. 2009;43(24):9525–30.
[50] Chandra R, Mohan SV. Microalgal community and their growth conditions influence biohydrogen production during integration of dark-fermentation and photo-fermentation processes. Int J Hydrogen Energy. 2011;36(19):12211–9.
[51] Chiu S-Y, Kao C-Y, Chen T-Y, Chang Y-B, Kuo C-M, Lin C-S. Cultivation of microalgal Chlorella for biomass and lipid production using wastewater as nutrient resource. Bioresour Technol. 2015;184:179–89.
[52] Vigani M, Parisi C, Rodríguez-Cerezo E, Barbosa MJ, Sijtsma L, Ploeg M, et al. Food and feed products from micro-algae: market opportunities and challenges for the EU. Trends Food Sci Technol. 2015;42(1):81–92.
[53] Seth JR, Wangikar PP. Challenges and opportunities for microalgae-mediated CO_2 capture and biorefinery. Biotechnol Bioeng. 2015;112(7):1281–96.
[54] Beal CM, Gerber LN, Sills DL, Huntley ME, Machesky SC, Walsh MJ, et al. Algal biofuel production for fuels and feed in a 100-ha facility: A comprehensive techno-economic analysis and life cycle assessment. Algal Res. 2015;10:266–79.
[55] Tercero EAR, Domenicali G, Bertucco A. Autotrophic production of biodiesel from microalgae: an updated process and economic analysis. Energy. 2014;76:807–15.
[56] Kouzuma A, Watanabe K. Exploring the potential of algae/bacteria interactions. Curr Opin Biotechnol. 2015;33:125–9.
[57] Ma X, Zhou W, Fu Z, Cheng Y, Min M, Liu Y, et al. Effect of wastewater-borne bacteria on algal growth and nutrients removal in wastewater-based algae cultivation system. Bioresour Technol. 2014;167:8–13.
[58] Kim D-H, Jang S, Yun Y-M, Lee M-K, Moon C, Kang W-S, et al. Effect of acid-pretreatment on hydrogen fermentation of food waste: microbial community analysis by next generation sequencing. Int J Hydrogen Energy. 2014;39(29): 16302–9.
[59] Sivagurunathan P, Sen B, Lin C-Y. Batch fermentative hydrogen production by enriched mixed culture: combination strategy and their microbial composition. J Biosci Bioeng. 2014;117(2):222–8.
[60] Hwang J-H, Kabra AN, Kim JR, Jeon B-H. Photoheterotrophic microalgal hydrogen production using acetate-and butyrate-rich wastewater effluent. Energy. 2014;78:887–94.
[61] Lakatos G, Deák Z, Vass I, Rétfalvi T, Rozgonyi S, Rákhely G, et al. Bacterial symbionts enhance photo-fermentative hydrogen evolution of Chlamydomonas algae. Green Chem. 2014;16(11):4716–27.
[62] Krohn-Molt I, Wemheuer B, Alawi M, Poehlein A, Güllert S, Schmeisser C, et al. Metagenome survey of a multispecies and alga-associated biofilm revealed key elements of bacterial-algal interactions in photobioreactors. Appl Environ Microbiol. 2013;79(20):6196–206.

[63] Boboescu IZ, Gherman VD, Ilie M, Lakatos G, Pap B, Negrea A, et al. Converting organic-rich industrial wastewater into biohydrogen through a novel hybrid dark fermentation–photoheterotrophic approach: proof of concept. J Biotechnol. 2015;(208):S33.
[64] Kumar V, Bhalla A, Rathore AS. Design of experiments applications in bioprocessing: concepts and approach. Biotechnol Prog. 2014;30(1):86–99.
[65] Boboescu IZ, Ilie M, Gherman VD, Mirel I, Pap B, Negrea A, et al. Revealing the factors influencing a fermentative biohydrogen production process using industrial wastewater as fermentation substrate. Biotechnol Biofuels. 2014;7(1):139.
[66] Lee D-H, Chiu L-H. Development of a biohydrogen economy in the United States, China, Japan, and India: With discussion of a chicken-and-egg debate. Int J Hydrogen Energy. 2012;37(20):15736–45.

Chapter | eight

Energy and Safety of Hydrogen Storage

Federico Ustolin[a], Jacob J. Lamb[b], Odne S. Burheim[c], Bruno G. Pollet[c]

[a]Department of Mechanical and Industrial Engineering, Faculty of Engineering, NTNU, Trondheim, Norway; [b]Department of Electronic Systems, Department of Energy and Process Engineering, ENERSENSE, Faculty of Information Technology and Electrical Engineering, NTNU, Trondheim, Norway; [c]Department of Energy and Process Engineering, ENERSENSE & NTNU Team Hydrogen, Faculty of Engineering, NTNU, Trondheim, Norway

CHAPTER OUTLINE

ABSTRACT

Hydrogen is considered a clean fuel and is required to replace fossil fuels in order to reduce environmental pollution. It can be transported through a pressurized tank in a gaseous phase or by means of cryogenic vessels in a liquid phase. The

Hydrogen, Biomass and Bioenergy. http://dx.doi.org/10.1016/B978-0-08-102629-8.00008-6

energy required by the compression and liquefaction processes, as well as the safety requirements, are considerably different. This chapter aims to compare the energy efficiency and the safety aspects of these two storage and transportation methods. A parametric numerical simulation model was developed to estimate the whole energy consumption of compressed gaseous hydrogen (CGH_2) tube trailer and liquid hydrogen (LH_2) road tanker. Regarding the safety aspects during storage and transport, the critical events and the consequences of the main accident scenarios were considered.

INTRODUCTION

Hydrogen (H_2) is considered a clean fuel and could replace fossil fuels in order to reduce environmental pollution. Potentially, its combustion produces only water and heat if the flame temperature is controlled or a catalyst burner is adopted with an appropriate H_2/O_2 ratio [1]. Moreover, hydrogen has a specific energy value (119.93 MJ/kg [2]) higher than other commercial fuels such as gasoline (43.45 MJ/kg [2]) or natural gas (47.14 MJ/kg [2]). Furthermore, hydrogen is not toxic and corrosive, and is harmless for humans and the environment [3].

Despite these and other advantages, H_2 is considered a dangerous fuel mainly due to its flammability range (4%–75% vol [4]) and low-ignition energy (0.017 mJ in air [5]). It is colorless and odorless, therefore challenging to detect in case of leakages. H_2 is also difficult to contain since it is the smallest and lightest element in nature, resulting in leaking of H_2 in some cases. H_2 has, in fact, a low density at standard temperature and pressure (0.0838 kg/m^3 at 293 K, 101.3 kPa bar [6]) compared with other fuels. Therefore, it is buoyant in air, and it can be dispersed easily in the atmosphere if leakage occurs in an unconfined environment. Contrarily, if the leakage happens in a confined environment, an explosive atmosphere will be rapidly created. Hence, when the transportation and utilization of H_2 are considered as appropriate options, the safety aspects should be accurately evaluated and taken into account.

Liquefaction increases the H_2 density (70.9 kg/m^3 at 20.4 K, 101.3 kPa [7]) and this is why liquid H_2 (LH_2) can be considered both for storage and transportation. In the case of road transportation, a truck with a tube trailer of compressed gaseous H_2 can be filled with 300–400 kg of H_2 at 200–250 bar, while a truck with a vacuum-insulated tank can hold up to 3.5 tons of LH_2 [8]. On the other hand, the energy requested for the liquefaction is higher than for the compression. Furthermore, the initial investment should be considered when a liquefaction plant is compared with a traditional (compressed) H_2 facility. For these reasons, LH_2 is usually more expensive than CGH_2. When LH_2 is used, different hazards and related scenarios should also be considered. Some of these have not been fully understood or forecasted yet.

H_2 is already widely used in several applications such as the chemical industry (63%) and refineries (31%) for several processes [9]. Usually, H_2 production and utilization take place within the same industrial facility. This choice is driven by safety, logistic, and economic aspects [9]. In the last sixty years, H_2 has been employed in the aerospace industry both as a propellant [10] and to power fuel cell systems (e.g., Bacon fuel cell, or alkaline fuel cell; AFC) in order to produce electric energy [11]. For this purpose, H_2 is stored in the liquid phase. Both H_2 production and consumption are expected to grow in the next coming years [12] mainly thanks to the energy storage capability of H_2 coupled with renewable energies. In this case, H_2 costs can be reduced, increasing the energy efficiency of the whole H_2 life cycle. Moreover, the H_2 safety aspects would become more critical, and atypical accident scenarios that have not been considered yet for H_2 could arise.

HYDROGEN STORAGE

As described in the introduction, one of the primary drawbacks of H_2 is the low density. Compared with other fuels, H_2 requires a more substantial volume, if stored at atmospheric conditions. In most of the applications, the storage volume is a critical issue. For this reason, different H_2 storage methods have been studied in the last decades. H_2 can be compressed or liquefied in order to increase its volume. Other storage methods foresee to bond it with a porous material such as metal-organic framework (MOF) or zeolites through physical adsorption (and desorption).

Moreover, H_2 can be stored through a chemical bond (adsorption) within metal hydrides (MH) or liquid organic hydrogen carriers

(LOHCs). A review of H_2 storage can be found in [13]. In the following, only the compressed and liquid H_2 solutions have been described.

Compressed gaseous hydrogen (CGH_2)

Compressed H_2 is the most convenient and common H_2 storage technique. The compression can increase the H_2 density up to 39.2 kg/m^3 at 700 bar and room temperature [7]. Increasing the H_2 pressure, the energy request by the compression increases as well. Theoretically, 1.05 kWh/kg$_{H2}$ are requested to compress H_2 isothermally from 20 to 350 bar, while 1.36 kWh/kg$_{H2}$ are needed to compress it up to 700 bar [14]. Hence, the energy required for compression is not linear. In practice, 1.7–6.4 kWh/kg$_{H2}$ is absorbed by the compressor. Also, the equipment (e.g., compressors, tanks, pipes, and valves) initial costs increase with higher pressures. Nowadays, four types of H_2 tanks exist (Type I, II, III, and IV) [15]. These tanks are made of different materials. Usually metal, composite (carbon fibers and resin), and plastic are used. The different tank types can resist at different pressures. The choice of the materials should be accurate when H_2 is used to reduce the costs and satisfy the safety criteria.

Liquid hydrogen (LH_2)

LH_2 is a solution adopted for a large amount of H_2. LH_2 has a density of 70.9 kg/m^3 at 20.4 K and atmospheric pressure [7]. Due to its low-boiling temperature (−253°C), H_2 can quickly evaporate. That is why the LH_2 tanks are composed of two metal tanks separated by a vacuum jacket filled with insulating material [16]. Even if the LH_2 tank is very well-insulated, the heat losses with the atmosphere are sufficient to make the LH_2 evaporate. The boil-off gas (BOG) formation is one of the main drawbacks of the LH_2 storage method.

At atmospheric conditions, H_2 is composed of 75% of ortho- and 25% of para-H_2. During the liquefaction, H_2 is usually converted to 100% para-H_2 in order to reduce the BOG rate. This conversion is carried out through a catalyst, and it is an exothermic process, hence requires heat removal. The theoretical energy for the liquefaction from atmospheric conditions is 3.3 kWh/kg$_{H2}$ and 3.9 kWh/kg$_{H2}$ without and with the conversion from ortho- to para-LH_2, respectively. In practice, the energy requirements for the liquefaction are 10–13 kWh/kg_{H2} [14]. This amount of energy corresponds to 30%–40% of the H_2 lower heating value (LHV) that is 120 MJ/kg (33.3 kWh/kg). Although, part of the heat could be recovered during the evaporation of the LH_2, decreasing the total energy consumption. In this case, the selection of the material should be carried out, also considering the low-temperature effects. Despite all these drawbacks, the interest in this type of storage is growing thanks to the high-energy density compared with other H_2 storage solutions.

CASE STUDY OF ENERGY COMPARISON BETWEEN CGH_2 AND LH_2

H_2 compression is a widespread solution to store H_2 since it is cheaper than other methods and less consuming from an energy point of view at the same energy density. Instead, liquefaction is one of the most demanding storage solutions. The advantage of this method is the achievable H_2 density (70.9 kg/m^3 [7]). Only with cryo-compressed H_2 storage, the density can be higher (up to 80 kg/m^3 [17]). LH_2 storage is necessary for some applications, such as an aerospace rocket, in which a large amount of H_2 is used. This study aims to investigate when it is more convenient for storage and transport of compressed gaseous or LH_2. Two case studies have been analyzed. In the first one, H_2 is transported to a gas station after its production as a compressed gas (CGH_2), while in the second one, H_2 is in the liquid phase (LH_2). In the facility, the fuel is compressed again at a higher pressure in order to fill the H_2-fueled vehicles equipped with 700 bar tanks.

A script in MATLAB has been developed in order to estimate the energy consumption of all the phases needed to deliver H_2 after its production to a gas station. The phases are the following—H_2 compression or liquefaction processes, CGH_2 or LH_2 road transportation, and H_2 compression up to 880 bar. The script is parametric, the amount of H_2, and the delivery range have been chosen as variables. The mass of H_2 varies between 100 and 4000 kg, while the driving range between 0 and 2000 km. The best storage solution from an energy point of view has been estimated for each configuration.

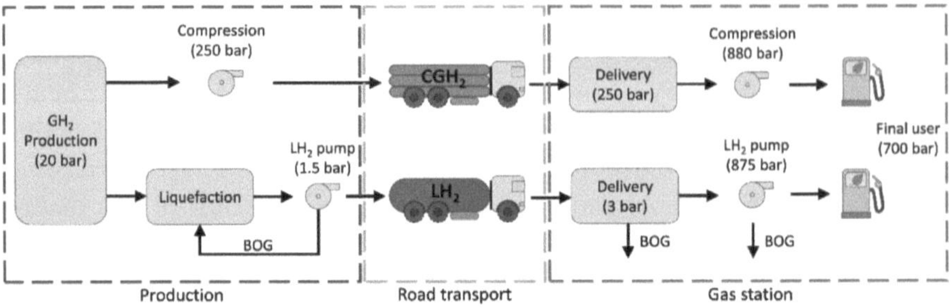

FIGURE 8.1 Schematic of the considered phases in the two case studies. The energy needed for GH_2 production has not been considered.

The schematic shown in Fig. 8.1 of the considered phases for the two case studies is shown. For both cases, H_2 is already compressed at 20 bar after its production [14]. The energy needed to produce H_2 has not been taken into account in the comparison since it is the same for the two case studies. The masses of the CGH_2 tube trailer and the LH_2 road tanker are the same, equals to 40000 kg [18].

In the case of CGH_2, H_2 is transported at 250 bar through a tube trailer with four type IV tanks. This type of vessel is made of composites materials, a polymer or a thin metallic liner wrapped with carbon fibers embedded in a polymer matrix [15]. These vessels have a cylinder shape and are transported with a truck, so-called tube trailer. The TITAN module (4-tube ISO container) has been selected as CGH_2 tube trailer. It contains 616 kg of H_2 at 250 bar, of which 585 kg are deliverable with an efficiency of 94% [19]. Tanks capable of bearing up to 700 bar [18] already exist, but regulations limit the tube trailer to a pressure of 250 bar for road transportation [20]. One of the targets set by the US DoE (US Department of Energy) for 2020 is to reach 520 bar of filling pressure [21], [22]. Hence, the first compression of the first case study goes from 20 up to 250 bar. At the gas station, H_2 is delivered to the facility and compressed again when required in order to fill the vehicles at 700 bar. The overpressures needed to fast fill 350 and 700 bar tanks are 440 and 880 bar, respectively [14]. Thus, the second compression reaches 880 bar.

In the second case study, the liquefaction takes place after H_2 production and is the most energy demanding process. Afterward, the road tanker is filled at a pressure of 1.5 bar [23]. The LH_2 tank used for road transportation has a volume of 17,000 gallons (64.4 m^3) [24]. The maximum H_2 capacity for this type of tank is 3810 kg_{H2} that corresponds to a deliverable H_2 mass of 3650 kg due to the boil-off formation [23]. At the gas station, LH_2 is transferred from the road tanker into the Dewar (insulated-cryogenic container) of the facility. When H_2 is needed by the vehicles, it is compressed up to 875 bar through an LH_2 pump.

In the next sections, the boil-off gas (BOG) losses and the energy consumption values for each phase have been reported and described. In Table 8.1, the truck parameters have been collected.

Boil-off losses

H_2 tends to evaporate when it is in liquid phase due to the vast temperature difference between LH_2 (20.4 K at 1 bar) and ambient air. For this reason, the cryogenic vessel should always be very well-insulated in order to reduce the radiation and convective heat flux from the environment.

During the filling operation, 3.3% of the total mass of H_2 transferred from the liquefaction facility to the tank is vented [23] due to the BOG formation. This amount of H_2 has not been considered as wasted since it is usually recycled back to the liquefaction plant [23].

TABLE 8.1 Truck And Tank Parameters Assumed in the Case Studied

Parameter	Ref	Value
CGH_2 and LH_2 trucks mass	[18]	40,000 kg
CGH_2 tube trailer payload	[19]	616 kg
CGH_2 tank deliverable efficiency	[19]	95%
CGH_2 tube trailer pressure	[20]	250 bar
LH_2 road tanker payload	[23]	3,810 kg
LH_2 road tanker deliverable mass	[23]	3,650 kg

According to the e-CFR regulations, the insulation of the LH_2 road tanker should be designed in order to avoid the venting for at least 110 h [25]. This time is lower than the one needed by the road tanker to cover 2000 km, that is, the maximum considered range. For this reason, BOG formation has not been considered during the LH_2 road transport.

BOG formation has been accounted for as a loss when LH_2 is transferred from the road tanker to the gas station Dewar since the vented gas is dispersed in the atmosphere and cannot be recovered. When H_2 flows through the LH_2 pump, BOG is formed and considered as wasted again. The amount of H_2 vented due to the delivery to the Dewar and pump losses varies with the size of the facility. For small commercial fueling stations ($< 1000\ kg_{H2}$/day), the BOG losses are 15.4% of the pumped LH_2 while for large fueling stations ($> 1{,}000\ kg_{H2}$/day) these losses are estimated for 5.9% of the pumped fuel [23]. In the developed script, the BOG losses have been estimated for each simulated configuration of the second case study (LH_2 method). Then, the wasted H_2 mass is added to the H_2 mass before the liquefaction. This implies a larger volume needed by the LH_2 road tanker at the beginning of the travel.

Energy consumption

In the following, the methodology used to estimate the energy required for the different processes, such as H_2 compression and liquefaction, have been described. In order to estimate the theoretical energy required by the H_2 compression, the equations reported in [26] have been used. The internal work needed for the compression is given by:

$$w_{i,\,comp} = h_2 - h_1 + q_k \tag{8.1}$$

Where h_1 and h_2 are the enthalpies before and after the compression, respectively, and q_k is the cooling energy given by:

$$q_k = T \cdot \Delta s \tag{8.2}$$

Where T is the temperature of the gas during the compression and Δs is the difference in entropy between the final and the starting point. The cooling energy is equal to the reversible heat in case of absence

TABLE 8.2 Enthalpy and Entropy Values for Different Thermodynamic States [7]

Pressure (bar)	Temperature (K)	Enthalpy (kJ/kg)	Entropy (kJ/kg*K)
20	300	3,967	41.05
250	300	4,091	30.58
880	300	4,547	25.25

of internal friction. The temperature is supposed to be constant (i.e., 300 K) and the entropy difference between the initial and final stage can be found on the H_2 T-s diagram. In Table 8.2, the enthalpy and entropy values for the different considered thermodynamic states have been collected.

For example, the compression from 20 to 250 bar, the first one of the CGH_2 case study, requires the following ideal energy:

$$w_i = 4091 - 3917 + (300(41.05 - 30.58)) = 3265\, kJ/kg_{H2} = 0.91\, kWh/kg_{H2} \qquad (8.3)$$

According to [14], an appropriate value of the compressor isentropic efficiency is 56% while the motor efficiency is 92% giving a total compression efficiency of 52%. This value is confirmed in [26] where a compression efficiency of 50% has been considered. Hence, the energy needed by the previous compression becomes 6337 kJ/kg_{H2}, equal to 1.76 kWh/kg_{H2}. When H_2 is compressed up to 700 bar, the compression efficiency decreases to 49% [14].

The second compression of the first case study goes from 250 bar (tube-trailer pressure) up to 880 bar, pressure needed to fast fill the 700 bar tank. Applying the previous equations for these two different pressure levels, the requested energy results in 4194 kJ/kg_{H2}, equal to 1.16 kWh/kg_{H2}. In this case, the overall compression efficiency, 49% and 0.2 kWh/kg_{H2} needed to cool down the gas to -40°C [14], has been added to the compression requested energy. This additional cooling energy is not needed for the first compression up to 250 bar.

The whole energy required for the two compressions is 3.12 kWh/kg_{H2}. New technologies, such as the ionic liquid piston compressor, are able to compress H_2 from 5 to 1000 bar consuming only 2.7 kWh/kg [27]. In this study, this technology has not been taken into account since it is still more expensive than traditional reciprocating compressors and not widespread yet.

LIQUEFACTION

The minimum ideal work needed by the liquefaction can be estimated by [26]:

$$w_{i,\,liq} = T_0 \cdot \Delta s + q_k \tag{8.4}$$

Where T_0 is the temperature of the gas at the beginning of the liquefaction, Δs is the difference in entropy between the final and the starting point and q_k is always the cooling load and in this case is given by:

$$q_k = h_2 - h_1 \tag{8.5}$$

In Table 8.3, the thermodynamic values for the considered states for the liquefaction have been collected. For example, if the liquefaction is carried out from 300 K and 1 bar down to the H_2 boiling point (20.3 K), the ideal energy required is:

$$w_{i,\,liq} = 300 \cdot (53.54 - (-0.02)) + (-0.45) + 3{,}958 = 12{,}103\, kJ/kg_{H2} \tag{8.6}$$

Increasing the H_2 pressure decreases the energy requested by the liquefaction. If H_2 has a pressure of 20 bar, 8,381 kJ/kg$_{H2}$ (2.3 kWh/kg$_{H2}$)

TABLE 8.3 Enthalpy And Entropy Values for Different Thermodynamic States [7]

Pressure (bar)	Temperature (K)	Enthalpy (kJ/kg)	Entropy (kJ/kg*K)
1	20.3	−0.45	−0.02
1	273	3,572	52.17
1	300	3,958	53.54
20	273	3,580	39.79
20	300	3,967	41.16

is the minimum theoretical work for the liquefaction process. Overall efficiency of 30% is typical for the liquefaction plants [26]. Furthermore, 0.65 kWh/kg_{H2} (2,340 kJ/kg_{H2}) are needed for the conversion from normal H_2 (75% ortho-, 25% para-H_2) to para-H_2 [14]. Hence, the total requested energy for the liquefaction starting from 300 K, and 20 bar is:

$$w_{liq} = \frac{8,381 + 2,340}{0.3} = 35,737\, kJ/kg_{H2} = 9.9\, kWh/kg_{H2} \tag{8.7}$$

In [28], large-scale H_2 liquefaction processes have been optimized, achieving an overall energy consumption of 7 kWh/kg_{H2}. However, 10 kWh/kg_{H2} has been assumed for this process since it is a more realistic value for the liquefaction [16]. This amount of energy corresponds to almost 30% of the H_2 LHV that is 33 kWh/kg [29].

ROAD TRANSPORT

As mentioned earlier, the CGH_2 tube trailer and the LH_2 cryogenic road tanker, have the same mass equal to 40,000 kg at full load [18]. It has been assumed that the fuel consumption changes linearly with the load rate. The trucks are fueled by Diesel and are powered by internal combustion engines (ICE). The LHV of Diesel is 42.96 MJ/kg (11.93 kWh/kg) with a density of 820 kg/m^3 [30]. The same fuel consumption of 40.5 L/100km at 60 km/h, the average speed of the trucks, has been considered for both vehicle types [31]. The amount of energy requested by the trucks is calculated as:

$$En_{truck} = \frac{C_{truck} \cdot \rho \cdot d \cdot LHV}{10^5} = [MJ] \tag{8.8}$$

Where C_{truck} is the fuel consumption in l/100 km, ρ is the density of Diesel, d the driving range, and LHV the Diesel energy content in MJ/kg. For example, if the truck should cover a distance of 100 km, the energy needed is:

$$En_{truck} = \frac{40.5 \cdot 820 \cdot 100 \cdot 42.96}{10^5} = 1,427\, MJ = 396\, kWh \tag{8.9}$$

The energy required by road transportation becomes significant when the delivery distance increases and the load of the truck decreases. The maximum amount of the deliverable LH_2 for one road tanker (3650 kg_{H2}) corresponds to almost 10% of the truck mass, while the maximum CGH_2 mass (616 kg_{H2}) is approximately 1.5% of the truck mass. Thus, the fuel consumption is not constant for all values of H_2 mass. Even if the tank load rate does not have a significant influence on fuel consumption, especially in the case of CGH_2, its effect has been considered in the developed script. Therefore, if the H_2 mass is 100 kg, the fuel consumption for the road trucker will be 91.1% of 396 kWh and 98.7% for the CGH_2 tube trailer.

HIGH-PRESSURE GH_2 FROM LH_2

After the delivery of the LH_2 to the gas station, H_2 must be changed in the gaseous phase and compressed in order to fill the vehicles' tanks. To carry out this process, an LH_2 cryogenic pump is used. LH_2 at 3 bar flows from the Dewar through the pump and is delivered as cryogenic compressed gas up to 875 bar between 30 and 60 K [32]. This pressure level is closed to 880 bar, the pressure required to fill the 700 bar tanks fast. Therefore, this requirement is considered satisfied. The pump requires an average energy of 1.1 kWh/kg_{H2} during the filling operation [32].

Results and discussion

In Table 8.4, the values of the energy required by the different processes have been collected together with the assumed efficiency for each phase. As mentioned earlier, the energy required by the liquefaction is always higher than the compression energy. Instead, the energy required by road transportation varies significantly with the driving range and the number of trucks. For example, in the considered case studies, this energy can reach a maximum value of 55,372 kWh when seven trucks are needed to transport 4000 kg of CGH_2 for a distance of 2000 km.

In Fig. 8.2, the 3-D chart shows the overall energy required by the CGH_2 and LH_2 solution when the H_2 mass and the delivery range vary. The *red* surface represents the CGH_2 solution, while the LH_2 storage is shown in *green*. These surfaces are not smooth but made of different steps representing the number of trucks needed for each H_2 mass value. Up to seven trucks are needed for the CGH_2 case study and only two for the LH_2 solution. If the transportation is not considered (0 km case), the

TABLE 8.4 The Energy Required by Different Phases

Process Phase	Ref.	Efficiency	Value
Compression from 20 to 250 bar	[14,26]	52%	6,337 kJ/kg_{H2} 1.76 kWh/kg_{H2}
Liquefaction process	[14]	30%	3,600 kJ/kg_{H2} 10 kWh/kg_{H2}
Truck trailer fuel consumption	[31]	/	40.5 l_{diesel}/100km
Truck energy consumption	/	/	1426 MJ/100km 396 kWh/100km
Compression from 250 to 880 bar	[14,26]	49%	4,896 kJ/kg_{H2} 1.36 kWh/kg_{H2}
Compression of LH_2 up to 875 bar	[32]	/	3,960 kJ/kg_{H2} 1.10 kWh/kg_{H2}

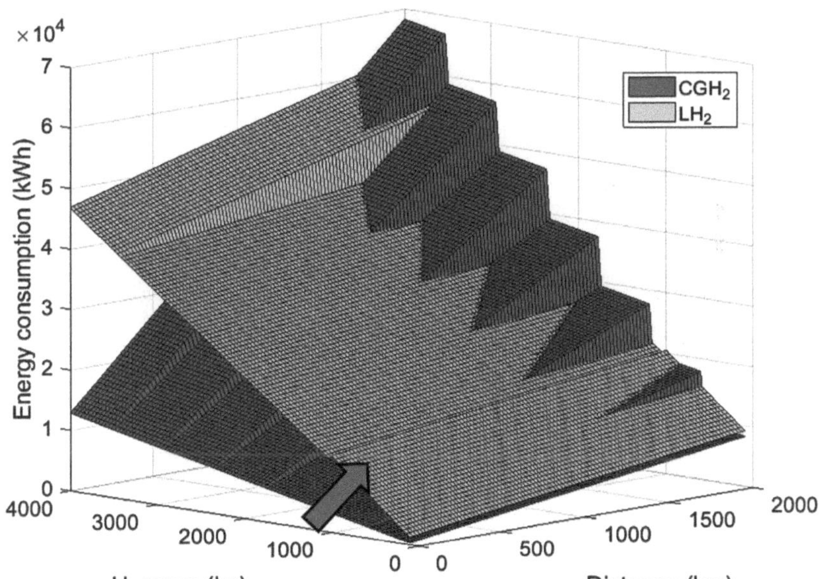

FIGURE 8.2 Energy consumption of the overall processes studied. In *red* and in *green*, the CGH_2 and the LH_2 process, respectively.

energy needed for the LH_2 case is always higher than the CGH_2 since the liquefaction process is more demanding than the compression. The same situation occurs when the H_2 mass is lower than 585 kg, despite the fact that the distance increases up to 2,000 km. For this H_2 mass range, the energy required by the two types of trucks is similar since only one truck is needed by both solutions. In Fig. 8.2, the *red arrow* shows a small bend in the LH_2 surface caused by the changing in the BOG rate as described in the boil-off losses section earlier.

To better compare the two case studies, the energy consumption matrix has been plotted and is shown in Fig. 8.3. The *green* surface represents the LH_2 consumption again while the *yellow* surface indicates the other solution. It means that when the LH_2 surface is shown on the graph, the CGH_2 method is the most convenient and vice versa. This is a 100 x 100 matrix with a resolution of 20 km and 40 kg of H_2. In Fig. 8.3, the *red* circles indicate the need for an additional truck by the LH_2 solution, while for each sawtooth-shaped step, one

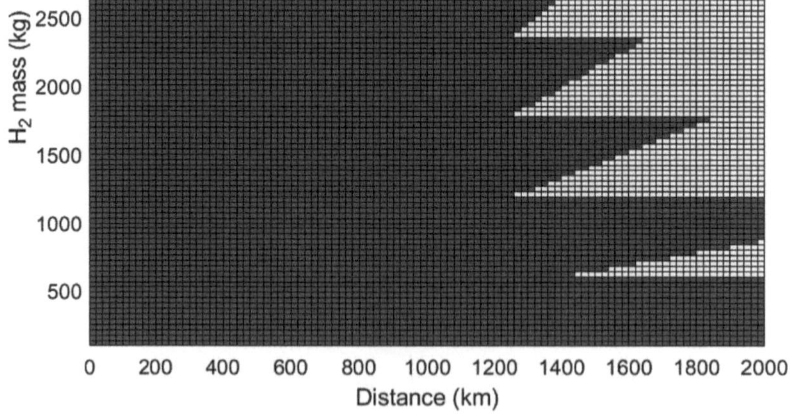

FIGURE 8.3 Energy comparison between the CGH_2 and LH_2 case studies as a function of H_2 transported mass and delivery distance.

tube trailer is added for the CGH_2 delivery. LH_2 energy consumption is always the highest when the delivery range is lower than 1260 km. On the other hand, if the H_2 mass is higher than 1192 kg, the LH_2 solution is always the most convenient when the driving range is higher than 1880 km.

From the obtained results, it seems clear that LH_2 is convenient when a large amount of H_2 (>1,200 kg) is transported, and the driving range is higher than at least 1260 km. In [33], LH_2 method is considered to be more convenient than CGH_2 when the demands are higher than 500 kg/day with mid-range distances. A similar comparison has been carried out from an economic point of view by Yang [34]. In that study also, the pipelines have been considered an alternative solution. The parameters chosen for the truck were slightly different, the CGH_2 tube trailer payload was 300 kg of H_2 (244 kg deliverable) at 160 bar, while for the cryogenic road tanker a payload of 4000 kg has been set. Moreover, the highest H_2 pressure needed to refill the vehicles was 350 instead of 700 bar. The results showed that for an H_2 mass higher than 2 ton and a driving range longer than 325 km, the LH_2 method was cheaper than the CGH_2. It is difficult to compare an economic evaluation with an energy one. Substituting the parameters set in the economic study in the developed script, similar results are achieved, and the LH_2 solution starts to be energetically more convenient for driving range longer than 500 km and H_2 mass heavier than 2000 kg. Yang [34] also estimated that for the amount of H_2 higher than 80 ton, the pipelines solution was even better than the LH_2. Even if the LH_2 road tanker can transport up to 4000 kg, this remains a limitation when a greater H_2 mass is transported. In this case, different solutions, such as rail or ship tankers, should be investigated and considered in order to reduce the required energy over H_2 mass ratio.

A comment can be added, making note that the values of the parameters in [34] have been chosen according to the regulations of that time. The deliverable H_2 mass with a CGH_2 tube trailer increased from 243 kg at 160 bar in 2007 to 585 kg (+100%) at 250 bar today, while the LH_2 road tanker payload remained the same. Another parameter that should be improved is the energy required by the H_2 liquefaction that is the most demanding process. The same value considered in this study (10 kWh/kg_{H2}) was assumed by Peschka in 1992 [16].

RISK ASSESSMENT OF HYDROGEN TRANSPORTATION AND STORAGE

H_2 properties have been already discussed in the Introduction. It is understandable why the safety aspects should always be considered when H_2 is both stored and transported. Accidents can be provoked by several causes, and these are not discussed here. The main critical events that can occur to CGH_2 and LH_2 tanks during these types of operations are the start of a fire, a breach on the shell, a leak from the tank or a pipe, and the catastrophic rupture of the vessel [35]. These critical events usually lead to a loss of containment (LOC). According to [36], a LOC is an unplanned or uncontrolled release of material from primary containment, including nontoxic and nonflammable materials. Hence, leakages and spills are both LOCs. Therefore, there are several consequences related to LOCs. In Table 8.5, the consequences of critical events for pressurized and cryogenic vessels have been reported.

TABLE 8.5 Consequences of Accident Scenarios That Involve CGH_2 and LH_2 Tanks [35]

Consequence	Type of tank
Fire Gas jet (ignited) Gas dispersion Gas puff (ignited) Vapor cloud explosion (VCE) Flashfire Jet fire Missile ejection Overpressure generation Fireball	CGH_2 and LH_2
Pool formation Pool ignited Pool dispersion Two-phase jet Aerosol puff Aerosol puff ignited	LH_2

Looking at Table 8.5, one can conclude that CGH_2 is safer than LH_2 since fewer consequences can happen. In a risk assessment, the probabilities of reaching each critical event and the yield of its consequences are estimated. Thus, any kind of conclusion cannot be drawn. Toxic cloud and environmental damage are two consequences not included in a H_2 risk assessment since it is neither toxic nor corrosive. This is another aspect that makes H_2 more suitable than other fuels from an environmental and health point of view.

In the past, different risk assessments on H_2 have been carried out. The problem of these studies is that different approaches have been employed, making them difficult to integrate in a uniform guideline for designing H_2 systems [37]. Moradi and Growth [37] suggested the consultation of different databases such as Hydrogen Incident and Accident Database (HIAD), ARIA, or to utilize the HyRAM tool in order to improve the quality of the risk assessment.

During the IDEALHY project [38], a qualitative risk assessment of H_2 liquefaction, storage, and transportation has been carried out. This study highlighted a distinct lack in experimental data and models for LH_2 release and explosions that are problems that impede the study of a full quantitative risk analysis. Boiling liquid expanding vapor explosion (BLEVE) has been identified as the highest consequence phenomena but with the lowest probability to occur. This is a physical explosion that can happen after a catastrophic rupture of the cryogenic tank. Despite the fact that three BLEVE accidents occurred in the past for LH_2, there are no experimental results or models available. This is an example of the existing gap in LH_2 safety knowledge. The authors concluded by saying that higher risk is present when the road transportation and storage are employed in urban areas compared to a liquefier plant.

In [39], a Bayesian network approach has been employed to compare compressed and liquefied H_2 transportation and tank station risks. One of the results of this study is that pipeline and liquefied H_2 tank station present more substantial risk than compressed H_2 tank station and truck transport. The risk is higher for the LH_2 system due to the possibility of a large VCE, even if the probability is minimal. In this evaluation, the type of CGH_2 tank has not been considered. The material chosen for the vessel should influence the risk assessment. This is the case of composite H_2 storage vessels, which have less resistance to fire than traditional H_2 metal tank [40]. An appropriate and extensive risk assessment should be carried out in order to understand which is the safest solution for this specific application since it is not clear from this literature review.

CONCLUSIONS

In this chapter, an overview of the most common H_2 storage solutions has been given. Furthermore, an energy comparison between the CGH_2 and the LH_2 methods has been carried out. The results confirm previous studies in which the CGH_2 solution has been selected as the most suitable for a small amount of H_2 and short distance delivery. In particular, with the technologies available today, the LH_2 solution becomes more convenient in energy terms when the H_2 mass is larger than 1200 kg and the driving range longer than 1300 km. After all, the results have been compared with a similar study and seem in accordance when the same parameters are set.

A brief literature review has been carried out in order to investigate the safety aspects of H_2 storage and transport solutions. Few studies have been carried out on this topic, but they are difficult to compare and useless as guidelines since different approaches have been employed. The main accident consequences of the CGH_2 and LH_2 solutions have been compared. Furthermore, a previous safety comparison between these two methods has been briefly described in which the Bayesian network approach has been employed, and the CGH_2 solution seems the safest one.

Further improvements are needed for both solutions in order to decrease the final H_2 cost, especially in the case of the LH_2 method, which has an excellent potential for transportation of large H_2 amount. Moreover, other LH_2 transportation means, such as rail or ship tankers, should be investigated in order to cover the limitation of road transportation methods.

REFERENCES

[1] Fumey B, Buetler T, Vogt UF. Ultra-low NOx emissions from catalytic hydrogen combustion. Appl Energy 2018;213:334–42.

[2] Wang MQ. Greenhouse gases, regulated emissions, and energy use in transportation. Argonne National Laboratory. [Argonne has also released version 2.8a that includes the impact of vehicle manufacturing.] Available from: http://www.transportation.anl.gov/software/GREET/; 2008.
[3] NASA. Safety standard for hydgogen and hydrogen systems. National Aeronautics and Space Administration. 2005.
[4] Edeskuty F, Stewart W. Safety in the Handling of Cryogenic Fluids. New York; Plenum Press; 1996.
[5] Ono R, Nifuku M, Fujiwara S, Horiguchi S, Oda T. Minimum ignition energy of hydrogen–air mixture: effects of humidity and spark duration. J Electrostat 2007;65(2): 87–93.
[6] McCarty R, Hord J, Roder H. Selected Properties of Hydrogen (Engineering Design Data). NBS Monogr. Boulder, Colorado: National Bureau of Standards; 1981.
[7] Lemmon EW, McLinden MO, et al. NIST chemistry WebBook, NIST standard reference database, NIST Chemistry WebBook, 69; 2017. doi:10.18434/T4D303.
[8] Pritchard DK, Rattigan WM. Hazards of liquid hydrogen RR769 Position paper. 2010.
[9] Ausfelder F, Bazzanella A. Hydrogen in the chemical industry. In: Stolten PDD, Emonts DB, editors. Hydrogen science and engineering: materials, processes, systems and technology; 2016. doi:10.1002/9783527674268.ch02.
[10] Cecere D, Giacomazzi E, Ingenito A. A review on hydrogen industrial aerospace applications. Int J Hydrogen Energy 2014 Jul;39(20):10731–47.
[11] Baroutaji A, Wilberforce T, Ramadan M, Olabi AG. Comprehensive investigation on hydrogen and fuel cell technology in the aviation and aerospace sectors. Renew Sustain Energy Rev 2019;106:31–40.
[12] International Energy Agency. Global trends and outlook for hydrogen. Available from: https://ieahydrogen.org/pdfs/Global-Outlook-and-Trends-for-Hydrogen_Dec2017_WEB.aspx; 2017 [accessed 12.04.19].
[13] Burheim OS. Hydrogen for Energy Storage. In: Engineering Energy Storage. Cambridge, United States: Academic Press; 2017. p. 147–92.
[14] USDOE. Energy requirements for hydrogen gas compression and liquefaction as related to vehicle storage needs. Available from: https://www.hydrogen.energy.gov/pdfs/9013_energy_requirements_for_hydrogen_gas_compression.pdf; 2009 [accessed 12.04.19].
[15] Barthelemy H, Weber M, Barbier F. Hydrogen storage: recent improvements and industrial perspectives. Int J Hydrogen Energy 2017;42(11):7254–62.
[16] Peschka W. Liquid Hydrogen - Fuel of the Future. 1st ed. Wien: Springer-Verlag; 1992.
[17] Petitpas G, Moreno-Blanco J, Espinosa-Loza F, Aceves SM. Rapid high density cryogenic pressure vessel filling to 345 bar with a liquid hydrogen pump. Int J Hydrogen Energy 2018 Oct;43(42):19547–58.
[18] Antoniotti. Impact of high capacity CGH2-trailers, deliverable 6.4, FCH-JU-2009-1 Grant Agreement Number 278796. Available from: https://www.fch.europa.eu/sites/default/files/project_results_and_deliverables/Recommendations%20to%20industry%20%28ID%202849587%29.pdf; 2013 [accessed 12.04.19].

[19] Baldwin D, Hexagon Lincoln. Development of High Pressure Hydrogen Storage Tank for Storage and Gaseous Truck Delivery, Final Report. 2015.
[20] USDOE. Hydrogen tube trailers. Available from: https://www.energy.gov/eere/fuelcells/hydrogen-tube-trailers; 2019 [accessed 12.04.19].
[21] Gerboni R. Introduction to hydrogen transportation. Compend Hydrog Energy 2016 Jan;283–99.
[22] US Department of Energy. 3.2 Hydrogen Delivery, 2012. In Fuel Cell Technologies Office Multi-Year Research, Development, and Demonstration (MYRD&D) Plan. Annual Progress Report. 2012.
[23] Petitpas G. Simulation of boil-off losses during transfer at a LH2 based hydrogen refueling station. Int J Hydrogen Energy 2018;43(46):21451–63.
[24] Air Products and Chemicals Inc. Liquid hydrogen, Safetygram 9. Available from: https://www.airproducts.com/~/media/Files/PDF/company/safetygram-9.pdf; 2014 [accessed 12.04.19].
[25] Electronic Code of Federal Regulations. Title 49: Transportation, § 173.318 Cryogenic liquids in cargo tanks. Available from: https://www.govinfo.gov/app/details/CFR-2011-title49-vol2/CFR-2011-title49-vol2-sec173-318; 2019 [accessed: 12.04.19].
[26] Hirscher M. Handbook of Hydrogen Storage - New Materials for Future Energy Storage. WILEY-VCH: Germany; 2010.
[27] Sdanghi G, Maranzana G, Celzard A, Fierro V. Review of the current technologies and performances of hydrogen compression for stationary and automotive applications. Renew Sustain Energy Rev 2019;102:150–70. doi:10.1016/j.rser.2018.11.028.
[28] Cardella U, Decker L, Sundberg J, Klein H. Process optimization for large-scale hydrogen liquefaction. Int J Hydrogen Energy 2017 Apr;42(17):12339–54.
[29] Zittel W, Wurster R, Bolkow L, Advantages and disadvantages of hydrogen. Hydrogen in the energy sector, Systemtechnik Gmbitt, Sömmerda, Germany; 1996.
[30] Mollenhauer K, Tschoeke H. Hanbook of Diesel Engines. Springer-Verlag: Berlin/Heidelberg; 2010.
[31] The International Council on Clean Transportation. Fuel consumption standards for heavy-duty vehicles in India. Available from https://theicct.org/sites/default/files/publications/ICCT_India-HDV-fuel-consumption_policy-update_20171207.pdf; 2017 [accessed 12.04.2019].
[32] Petitpas G, Aceves SM. Liquid hydrogen pump performance and durability testing through repeated cryogenic vessel filling to 700 bar. Int J Hydrogen Energy 2018 Sep;43(39):18403–20.
[33] U.S. DRIVE. Hydrogen Delivery Technical Team Roadmap. Available from: https://www.energy.gov/sites/prod/files/2017/08/f36/hdtt_roadmap_July2017.pdf; 2017 [accessed 12.04.19].
[34] Yang C, Ogden J. Determining the lowest-cost hydrogen delivery mode. Int J Hydrogen Energy 2007 Feb;32(2):268–86.

[35] Delvosalle C, Fievez C, Pipart A, Debray B. ARAMIS project: a comprehensive methodology for the identification of reference accident scenarios in process industries. J Hazard Mater 2006;130(3 SPEC. ISS.):200–19.
[36] AIChE. Center for chemical process safety, loss of primary containment (LOPC). Available from: https://www.aiche.org/ccps/resources/glossary/process-safety-glossary/loss-primary-containment-lopc; 2019 [accessed 12.04.19].
[37] Moradi R, Groth KM. Hydrogen storage and delivery: review of the state of the art technologies and risk and reliability analysis. Int J Hydrogen Energy 2019;44(23):12254–69. doi:10.1016/j.ijhydene.2019.03.041.
[38] Hankinson G, Lowesmith BJ. Integrated design for demonstration of efficient liquefaction of hydrogen (IDEALHY), Qualitative Risk Assessment of Hydrogen Liquefaction, Storage and Transportation, Delivery Number 3.11. Available from: https://www.idealhy.eu/uploads/documents/IDEALHY_D3-11_Risk_Report_web.pdf; 2013 [accessed 12.04.19].
[39] Pasman HJ, Rogers WJ. Risk assessment by means of Bayesian networks: a comparative study of compressed and liquefied H_2 transportation and tank station risks. Int J Hydrogen Energy 2012;37(22):17415–25.
[40] Ruban S, Heudier L, Jamois D, Proust C, Bustamante-Valencia L, Jallais S, et al. Fire risk on high-pressure full composite cylinders for automotive applications. Int J Hydrogen Energy 2012 Nov;37(22):17630–8.

Chapter | nine

Future Prospects of Selected Hydrogen and Biomass Energy Technologies

Jacob J. Lamb[a], Bruno G. Pollet[b]

[a]*Department of Electronic Systems, Department of Energy and Process Engineering, ENERSENSE, Faculty of Information Technology and Electrical Engineering, NTNU, Trondheim, Norway;* [b]*Department of Energy and Process Engineering, ENERSENSE & NTNU Team Hydrogen, Faculty of Engineering, NTNU, Trondheim, Norway*

CHAPTER OUTLINE

ABSTRACT

Renewable energy sources are rapidly expanding. Although these sources of energy provide large quantities of low emission energy, due to the intermittent nature of these sources the focus of energy storage technologies has expanded. In particular, the accumulation of renewables has been deeply focused on water electrolysis, biomass gasification, and hydrocarbon reforming for hydrogen production. Here an overview of new research in these areas is presented. In addition, a comparison is made with fuel cell investigation and future trends in water electrolysis are discussed.

Hydrogen, Biomass and Bioenergy. http://dx.doi.org/10.1016/B978-0-08-102629-8.00009-8

OVERVIEW OF WATER ELECTROLYSIS RESEARCH TRENDS

The water electrolysis analysis areas are still in their early stages and contain several rising technologies. In microbial electrolysis cell (MEC) research, elementary studies of parts and practical analysis of the entire system are conducted at the same time. MECs will use versatile sources, as well as organic substrates like brine, irrigation, and wastewater, that have entirely different elements based on their origin. Moreover, the electricity required for electrolysis may be provided from renewable energy systems (RES) and sources, like solar or wind generation, that conjointly varies spatially. Therefore, MEC systems ought to be optimized case by case, and thus, research regarding the system can continue to expand.

Research is compatible with the catalysts used in their fuel cell alternatives [1,2] when considering alkaline electrolysis cells (AECs) and proton exchange membrane electrolysis cells (PEMECs), and many catalysts have not yet been transferred. In the past decade, AEC and PEMEC studies have gained considerable attention. This could be because in regions without adequate infrastructure, they are appropriate alternatives for on-site energy storage. AECs have long been researched and are well understood. PEMECs also have a lengthy history and extensive expertise that has accumulated. Focusing on AECs and PEME is therefore sensible for on-site energy storage systems.

Optimizing AEC and PEMEC processes are less critical than the MEC scheme because the variety of organic substrates need not be considered by these processes. Several techniques have not yet been implemented to AEC and PEMEC from fuel cell studies, as shown in the following fuel cell research reviews [1,2].

Both the the amount of solid oxide electrolysis cell (SOEC) publications and the amount of other water electrolysis technologies have risen recently; however, SOEC systems research is not as extensive as other types of research. This could be because SOEC's stability for long-term operation is inadequate. Research on SOEC generally involves studies on primary materials. Due to the difficult circumstances at the SOEC anode, which produce a large quantity of O_2 with elevated partial pressures at elevated temperatures, their stability is reduced (this is distinct from the solid oxide fuel cell scheme). This is, therefore, a novel challenge for SOEC that cannot be solved through the transfer of SOFC technology alone. Also, SOEC needs steady energy input to keep its elevated temperature, which is not appropriate for renewable energy storage on a low or medium scale. This is because the time dependence of renewable energy is not appropriate at these scales to maintain a steady temperature. This might also be the reason why SOEC systems research is not receiving much focus at the moment. If the stability of the anode becomes adequate and large-scale storage of renewable energy is needed, there will be more interest in SOEC studies.

Comparison of water electrolysis with fuel cell technologies

As of 2016, more than 94,000 fuel cell studies have been released, and nearly all techniques used for water electrolysis are also used for fuel cell research, indicating that water electrolysis research has examined techniques used in fuel cell research [1,2]. In fact, in fuel cell research, microbial fuel cells (MFCs), and catalysts for alkaline fuel cells (AFCs) and proton exchange membrane fuel cells (PEMFCs) are increasing the most. This model is comparable to water electrolysis studies as MEC's are the reverse techniques of MFC, and AEC and PEMEC catalysts are the same as in their fuel cell alternatives. This may be due to technology transferred from fuel cell research promoting water electrolysis systems growth. Future water electrolysis techniques could, therefore, be predicted from fuel cell research.

One of the most promising catalysts for water electrolysis is the metal-free N-doped graphene/carbon nanotube (CNT) catalyst, which shows elevated-oxygen reduction reaction (ORR) catalytic activity for low-temperature fuel cells. This could also be helpful for the response to oxygen evolution reaction (OER), as ORR is just the opposite of OER, so catalysts should operate the same way. The heteroatom-doped graphene/CNT can be replaced with boron (B), nitrogen (N), or phosphorus (P), providing distinct electronic status for carbon-based products [1,2]. Also explored were dual heteroatom (N and B/P/S)-doped graphene/CNT catalysts. Iron or nickel-based N-doped graphene/CNT catalysts are also emerging technologies in fuel cell studies, although they dissolve in acidic solutions and are used by AECs. These metal or platinum-free catalysts or non-precious metal catalysts (NMPC) may decrease the electrolyzer overall costs.

Interestingly, nickel phosphide has not been researched in fuel cell studies (which displays elevated catalytic activity for the OER and is stable in acid). Newly developed, Ni_5P_4 and CoP are primarily used in water electrolysis. These compounds may become more stable, allowing them to be used in fuel cells. This catalyst could be an instance of inverse transfer to fuel cell research from water electrolysis.

Nafion® is typically used in PEMECs as an electrolyte membrane and is adequate for water electrolysis because water availability maintains the proton (H^+) conductivity of the membrane, which reduces at low relative humidity. Electrolyte studies in PEMFC aim to enhance the conductivity of protons at low-relative humidity, which is pointless for water electrolysis since water electrolysis cells initially contain plenty of water [1,2]. Hydrocarbon-based polymers, however, contain hydrocarbons without the ecologically harmful fluorine and can be synthesized in a few steps, resulting in relatively low cost if the structure of the polymer is random, which differs from that of Nafion®. Although polymers with a random structure based on hydrocarbons do not display elevated conductivity of protons at low humidity, they may be adequate for water electrolysis below boiling point. In water electrolysis, hydrocarbon-based electrolytes are not often used and could be transmitted as inexpensive electrolytes. Also, steam electrolysis beyond boiling point has several benefits but experiences low proton conductivity at low humidity because steam is inadequate to humidify electrolytes beyond the boiling point at atmospheric pressure.

SOEC's anode stability is a challenge that is different from SOFC. The fuel cell research materials have been transmitted; however, the anode will degrade over long-term operation. SOFC research is nearly mature [1,2], meaning SOFC scientists and engineers will need to discover another subject linked to their skills that could be SOEC.

MEC research and MFC studies have extended rapidly. Much of the technology established in MFC research has not been applied to MEC. Direct electron transfer (DET) is of great interest in MFC [1–6]. The overpotential (η) induced by the transfer of an electron from the biocatalysts to the electrodes is a task in MFC. MFCs have traditionally used redox molecules or polymers capable of transferring electrons between the catalytic core and the electrode, although this decreases the MFC cell voltage. To prevent this, DET was explored using a redox protein in their membrane to transfer electrons from enzymes in a microbiome to electrodes. *Geobacter* and *Shewanella's* electron transfer mechanisms were explored in MFC for DET. This has been widely focused on MEC studies, but an inner loss could be reduced owing to redox molecules or polymers. MEC can prevent the production of O_2 in the anode, so it can be operated under anaerobic conditions that are appropriate for anaerobic bacteria such as *Geobacter* and *Shewanella*. MEC study can, therefore, use bacteria abandoned in MFC because of their sensitivity to O_2. The technology used in polymer electrolyte fuel cells (PEFCs) can solve the problems surrounding the electrolyte and platinum in the MEC cathode. The electrolytes evolved in PEFC studies and could be used in MEC with less expensive catalysts like N-doped graphene/CNT. In recent years, enzymatic biofuel fuel cells (EBFCs) were created. EBFC utilizes only biocatalyst enzymes and does not involve entire bacteria. It increases the current density of the device per unit volume. However, regenerating biocatalysts is hard, which is different from MFCs. The current density per volume is a low priority for water electrolysis, and EBFC technologies are seldom exported to MEC. The catalyst may be worth examining for use in MEC if a very stable catalyst is established in EBFC studies.

Biomass gasification

Biomass steam gasification is one of the primary thermochemical paths studied in the literature, resulting in an average H_2 output of 4 g (without catalyst) and 7 g (with catalyst) per 100 g biomass. As demonstrated by the low H_2 output achieved with the main catalyst, the need for a secondary catalyst in the gasification system is notable. This complicates the gasification process and, owing to the elevated temperatures required in both steps, raises energy expenses. Tar decrease is the main challenge of biomass gasification, as tar creates severe synthesis issues, even when secondary catalysts are used; therefore, it is essential to optimize the catalyst to enhance its stability and regeneration.

About the reforming of raw bio-oil steam, which was also researched widely in the literature, an average output of 10 g of H_2 per 100 g of biomass was accomplished [7]. The bio-oil's higher power density compared to biomass and the reduced temperatures required to reform it compared to biomass gasification make it an exciting strategy for bio-oil valuation, although the process's scaling is now its most significant challenge. Also, as stated above, bio-oil feeding still entails disadvantages as a portion of the bio-oil is lost in the steps of condensation and volatilization. Comparing the two approaches, biomass gasification is performed in big centralized plants situated in the areas where biomass is accessible, whereas biomass pyrolysis is performed in decentralized tiny pyrolysis plants in the bio-oil reforming approach and bio-oil is transferred to centralized bio-oil reforming plants, thereby decreasing transport expenses owing to higher energy expenses. However, biomass gasification technology is more advanced than bio-oil reforming technology, which is still in a preliminary stage.

Also, the findings achieved in biomass's two-step pyrolysis reform are promoting as an alternative way of biomass gasification and bio-oil reformation for H_2 manufacturing, reaching an average H_2 output of 10 g per 100 g biomass. This path enables both pyrolysis and reforming measures to be optimized and prevents raw bio-oil feeding issues. The primary benefit of this method is obtaining a free-tar product, which is the greatest biomass gasification challenge. Moreover, in the fields where biomass is accessible, it is an excellent approach, although there are still problems to be addressed. To increase this process, the deactivation of the catalyst and mainly its stability is crucial, although few literature studies deal with the deactivation of the catalyst and its impact on the distribution of the product. Furthermore, although distinct types of metal stages, supports, and promoters have been used, there are no studies in which their coke combustion or steam gasification regeneration has been performed, with these elements being essential for future studies.

The two-step method is a distinct way of producing $H_{2,}$ which can be an excellent solution for methods of gasification and bio-oil reforming [7]. Also, simple scaling is the primary benefit of the method of pyrolysis and in-line reforming, which in turn prevents the transport of the bio-oil required in the phase of reforming bio-oil. However, the use of one approach or another relies on the region in which it will be installed, their technological development, and the measures made to address the issues mentioned earlier.

Hydrocarbon reforming

While this study focuses on catalytic activity, the development of hydrocarbon reforming for H_2 or H_2-rich syngas generation for fuel cells has been the focus of a notable amount of study attempts to comprehend variables such as reforming catalyst composition, reforming response, and fuel processing. These regions are crucial as well but are not within this paper's scope. For fuel cells that are also highlighted, tolerance to impurities such as sulfur and anti-coking is essential. By conducting a comparative literature survey, it can be seen that the moist impregnation technique is the most prevalent method of preparing for the reforming catalyst. By using the suitable preparation technique and catalyst support, important catalyst parameters such as the conversion of hydrocarbons, manufacturing of H_2, and selectivity can be maximized. The addition of catalyst promoters and the selection of catalyst assistance can improve key catalyst parameters including H_2 selectivity, heat stability, chemical stability, and coal-deposition tolerance. The catalyst promoters are generally added to the Ni catalyst in the form of other metals, thus forming a bi-metallic alloy catalyst of M-Ni. It can be discovered from the literature that the excellent performing catalyst promoters are Co, Cu, Sn, Pt, Pd, Mn, Rh, Ru, and Au, which have been reported to significantly enhance the Ni catalyst's H_2 production while also reducing carbon deposition [8].

Growth in exploration in designing and identifying new reforming catalysts has resulted in the development of a new efficient catalyst with enhanced hydrocarbon reforming efficiency for H_2 or H_2-rich fuel cell gas generation. Sorption improvement and chemical looping steam reforming of hydrocarbons, especially methane, is promising to prevent or simplify the gas separation process. However, the major challenges are the cyclability of the CO_2 sorbent, the catalyst of the oxygen carrier, anti-coking, and slow kinetics. It is desired to develop new low-cost, chemical, and mechanically stable catalysts (keeping microstructure, negligible sintering/coarsening at high temperature), impurity-tolerant, anti-coking with high-catalytic activity to reform, and partial oxidation of hydrocarbons for H_2 production and fuel cell applications.

REFERENCES

[1] Ogawa T, Takeuchi M, Kajikawa Y. Analysis of trends and emerging technologies in water electrolysis research based on a computational method: a comparison with fuel cell research. Sustainability 2018;10(2):478.

[2] Ogawa T, Takeuchi M, Kajikawa Y. Comprehensive analysis of trends and emerging technologies in all types of fuel cells based on a computational method. Sustainability 2018;10(2):458.

[3] Lamb JJ, Hill RE, Eaton-Rye JJ, Hohmann-Marriott MF. Functional role of pila in iron acquisition in the cyanobacterium synechocystis sp PCC 6803. PLoS One 2014;9.(8).
[4] Lamb JJ. Characterization of bacterial electron transport to extracellular electron acceptors. Doctoral dissertation, Trondheim: NTNU. Available from: https://ntnu-open.ntnu.no/ntnu-xmlui/handle/11250/2399463; 2016.
[5] Lamb JJ, Hohmann-Marriott MF. Manganese acquisition is facilitated by PilA in the cyanobacterium Synechocystis sp. PCC 6803. PLoS One 2017;12(10).
[6] Lamb JJ. The role of Type IV Pili of Synechocystis Sp. PCC 6803 in extracellular electron transfer: a thesis submitted for the Degree of Master of Science. Dunedin, New Zealand: University of Otago; 2012.
[7] Arregi A, Amutio M, Lopez G, Bilbao J, Olazar M. Evaluation of thermochemical routes for hydrogen production from biomass: a review. Energy Convers Manag 2018;165:696–719.
[8] Sengodan S, Lan R, Humphreys J, Du D, Xu W, Wang H, et al. Advances in reforming and partial oxidation of hydrocarbons for hydrogen production and fuel cell applications. Renew Sustain Energy Rev 2018;82:761–80.

Target Audience

The general target audience for this Primer is industry professionals, researchers, and graduate students, especially those starting in a new topic of research or coming into the field.

Index

Note: Page numbers followed by "f" indicate figures, "t" indicate tables.

C

D

S

T

U

V

W

X

Z

Papier | Fördert gute Waldnutzung
FSC® C083411

Kolibri 360 GmbH
Ferdinand-Jühlke-Straße 7
99095 Erfurt, Deutschland
produktsicherheit@kolibri360.de